ENERGY DISSIPATION IN COMPOSITE MATERIALS

ENERGY DISSIPATION IN COMPOSITE MATERIALS

Peter A. Zinoviev
Yury N. Ermakov

CRC Press
Taylor & Francis Group
Boca Raton London New York

CRC Press is an imprint of the
Taylor & Francis Group, an **informa** business

Energy Dissipation in Composite Materials

First published 1994 by Technomic Publishing Company, Inc.

Published 2019 by CRC Press
Taylor & Francis Group
6000 Broken Sound Parkway NW, Suite 300
Boca Raton, FL 33487-2742

© 1994 by Taylor & Francis Group, LLC
CRC Press is an imprint of Taylor & Francis Group, an Informa business

First issued in paperback 2019

No claim to original U.S. Government works

ISBN-13: 978-0-367-44950-6 (pbk)
ISBN-13: 978-1-56676-082-9 (hbk)

Visit the Taylor & Francis Web site at
http://www.taylorandfrancis.com

and the CRC Press Web site at
http://www.crcpress.com

Main entry under title:
  Energy Dissipation in Composite Materials

Bibliography: p.

Library of Congress Catalog Card No. 93-61760

This book is about materials damping. All real materials, in one way or another, exhibit a departure from ideal elastic behavior, even at very small strain values. Under cyclic deformation, these departures result in irreversible energy losses in the material. The causes of such losses are many, and include the irreversible transfer of mechanical energy into heat, growth of cracks and other defects, and microplastic deformation of crystals, to name but a few. Several terms have been suggested to define these phenomena, including damping, energy dissipation, imperfect elasticity, and internal friction.

There are many wonderful things in the world which we never think about until they disappear or something goes wrong with them — things like air, health, love, and money. Damping is like that as well.

Imagine for a minute what the world would be like without damping. The noises of Egyptian chariots would still be heard in such a world, if they were not drowned out by the terrible thunder of Hiroshima. No machines would be operating, because the resonance of their work frequencies — in harmony with the natural vibration frequencies of their components — would shatter them. But the worst feature would simply be the lack of any static state, the absence of any rest, in a world of perpetual vibrations, oscillations, and noise. One can easily imagine the discomfort of such a world. Thank God it is only a fantasy.

It is not surprising that the dissipative behavior of materials has long attracted the attention of scientists and engineers. Unfortunately, most of the studies known to the authors describe different aspects of dissipation processes in common isotropic materials. These studies have only a limited application to modern technologies, which employ ever more complex anisotropic composite materials. This book is probably the first attempt to generalize, at the monograph level, a considerable number of different publications on energy dissipation in anisotropic bodies and composites.

Because it is the first attempt of its kind, the authors do not believe the selection of the material and its construction to be perfect. They will take with gratitude all comments about the subject of the book.

The Institute of Composite Technologies[1] sponsored the preparation of the manuscript. The authors are glad to express their deep appreciation to Dr. Ludmila P. Tairova, Director of ICT, for the thorough help in the work. Dr. Olga V. Lebedeva undertook the difficult but necessary work of translating the manuscript into the English language. The authors must inform the readers that the passages in the book (we hope they are not too numerous) which seem unclear in English, were still worse in Russian. We wish to express many thanks to Mr. Alexander I. Mochalov for his skillful typing and preparation of the text.

The idea of the book was actively supported by Mr. Michael Margotta, Vice President of Technomic Publishing Co., Inc. All practical work on intercontinental contacts with the authors and the final editing of the manuscript was performed by Ms. Susan G. Farmer. The authors will always remember with gratitude their contribution to the work, which was not only useful, but pleasant as well.

---

[1]ICT Ltd., is a nonstate scientific and commercial organization founded in 1991. It unites high-skilled specialists in the field of design and manufacture of structures made of advanced composite materials.

This book deals completely with damping or energy dissipation processes in vibrating solids. The property of material bodies which allows them to disperse energy under vibrations is often referred to as imperfect elasticity or internal friction, and has long been the subject of investigation. Experimental results and theoretical models describing damping effects are put forth in hundreds of papers and dozens of monographs. Some of them are given in the references list in the present book.

There are two reasons for our attention to these problems. The first one concerns internal problems of the theory of damping. As a rule, composite materials are anisotropic and nonuniform bodies, and description of damping processes in composites calls for essential new developments in the theory of damping. The methods for experimental investigation, the value and the meaning of damping characteristics, their relations with material internal structure, the rules to form and transform main relationships – these are just a few of the problems which need to be examined in order to fully describe energy dissipation processes in composite materials.

The second reason is a practical one – engineering. As a rule, modern composite materials dissipate energy during vibrations with much more intensity than do simple isotropic materials. With an understanding of these new characteristics, engineers will gain the ability to control damping characteristics of a wide range of materials by changing their internal structure, the orientation of reinforcing elements, the sequence and the types of the layers in multilayered materials. At the present time, engineers have already begun experimenting with these new techniques.

Figures I.1–I.8 show engineering systems for which their damping characteristics distinctly affect the quality of the system as a whole.

Figure I.1 depicts a space vehicle of the "Space Shuttle" type, with a two-

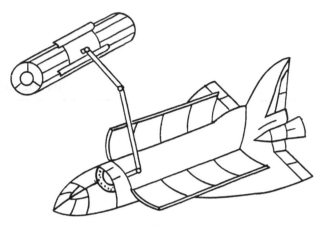

FIGURE I.1.

linked board manipulator. The manipulator is designed to eject different loads into space and to take them on board (for example, communication satellites). The duration (and with it the cost) of orbit operations, as well as the ultimate possibilities for precise manipulator operations, is directly related to the damping behavior of its links. This example is discussed in detail in Chapter 10. The damping response is also of extreme importance when it comes to designing the wings, the tail fins, and the other elements of modern planes (see Figure I.2). Dynamic aeroelasticity phenomena are the governing factors for such elements.

The processes of vibration damping are also very important in the ship-building industry. Two examples in Figures I.3 and I.4 illustrate this fact. The designers of comfortable cruise liners have to reckon with the fact that coddled passengers of deluxe class are unable to stand even the weakest

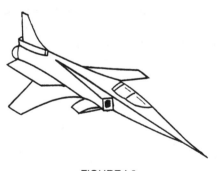

FIGURE I.2.

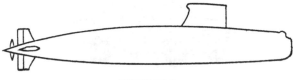

FIGURE I.3.

noise from the ship engine and shaft. This noise is of no less importance for submarine captains too. They worry, however, not about the crew comfort, but about the noise level that is transferred through the shaft and board elements into the environment.

Surface transport vehicles also need elements with high damping response. For example, a car's suspension (see Figure I.5) includes the stiffeners (the springs) and damping elements (the bumpers). It is quite possible to make the suspension system simpler and cheaper at the expense of increasing the spring damping ability, e.g., the springs can be made of glass fiber reinforced plastic.

Military systems are of course extremely sensitive to the quality of dynamic processes. For example, the accuracy of tank firing during its moving depends on how completely the barrel vibrations are damped (Figure I.6).

The problem of vibration damping is no less important for machine elements, e.g., for long and compliant mandrels (rods) of boring machines (Figure I.7).

Modern sports impose very high demands on sports equipment. Several examples of sports equipment for which the level of internal damping is of great importance are an archery bow, a vaulting pole, and a tennis racket (Figure I.8).

This book can be divided roughly into three parts. The first part, from Chapter 1 to Chapter 3, may be considered as the introductory section. It includes information on the main methods of approach to describing energy

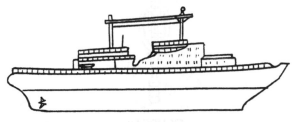

FIGURE I.4.

FIGURE I.5.

FIGURE I.6.

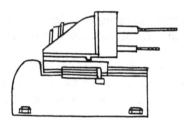

FIGURE I.7.

FIGURE I.8.

dissipation processes in materials (Chapter 1). Main methods for experimental investigation of the damping behavior are also described (Chapter 2) and experimental results are given for a set of materials (Chapter 3).

The second part of the book, from Chapter 4 to Chapter 8, contains basic results which describe the dissipative response of anisotropic bodies and composites. Chapter 4 includes general theoretical background for anisotropic bodies under a three-dimensional stress state. The dependence of dissipation parameters on stress or strain amplitudes is also considered. Chapter 5 deals with one of the major composite types – the unidirectional composite material or the monolayer. The structural model for elastodissipative characteristics of the monolayer is explained. Analogous operations are carried out in Chapter 6 within the limits of the viscoelastic model for dissipative behavior. Chapter 7 examines extreme properties of monolayer dissipation parameters. The extreme properties are of great importance for estimating the real range of dissipative properties which are accessible to the materials designer. Methods for graphically depicting the results have received much attention, as such methods significantly extend the information on material possibilities. Chapter 8 considers basic relations for multilayered hybrid composites, which enable engineers to relate monolayer properties and the laminate structure to the overall properties of the multilayered composite.

The third and the last part of the book covers Chapters 9 and 10. It treats the dissipative behavior of structural elements. To have a general idea of the rules for transfer from material dissipative properties to structure dissipative properties, it is sufficient, in the authors' opinion, to examine the simplest elements, viz., bars and beams. These are not only suitable practice examples, but they also enable engineers to analyze rather complicated practical problems as well, for example, the dissipative behavior of a two-linked space vehicle manipulator. Chapter 10 considers the problems of the optimal design of such structures.

One general idea which the authors tried to follow was the wish to make the book a complete one, so that the reader would not have to refer constantly to other publications. For example, dissipative properties indissolubly relate to elastic properties; the reader will be convinced of this fact. That is why the information on elastic properties is given in detail in the chapters where it is really required. For the same purpose there are the Appendices at the end of the book, which include necessary brief information from those branches of mathematics that are needed in understanding the book.

The authors hope that these efforts will make the book accessible for students and engineers.

CHAPTER 1

Main Methods of Approach to Describing the Phenomenon of Internal Friction in Solids

Cyclic loading of solids even under small strain amplitudes is accomplished by energy damping. This phenomenon is usually referred to as internal friction or imperfect elasticity. The energy dissipation factor ψ (also called the relative energy dissipation) may be considered as the parameter describing a system's internal friction. The dissipation factor for isotropic bodies is the ratio of energy losses, ΔW, in a loading cycle to the amplitude value of the elastic potential energy in a cycle, W:

$$\psi = \Delta W / W \qquad (1.1)$$

The dissipation factor, ψ, can depend upon the vibration frequency, ω, the temperature, T, the amplitude of vibrations, and other factors.

There is a set of theories and different methods of approach to describing energy dissipation processes in solid bodies. The present chapter considers briefly the most important of them.

1.1 SIMPLEST MODELS OF INTERNAL FRICTION

The simplest (engineering) models for describing internal friction are usually based on the analysis of some mechanical systems, the behavior of which is intuitively believed to be analogous to the behavior of the solids under loading.

Let us consider, for example, *the model for a body with "dry" (constant) friction*. The system shown in Figure 1.1 is the mechanical analog of such a body. The system consists of a load moving along a rough bar, and a spring connecting the bar and the fixed base. The constant force of "dry" friction affects the load. Only the direction of the load motion (the velocity

1

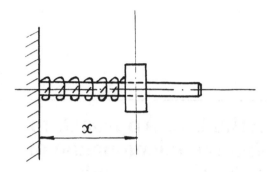

FIGURE 1.1. Model of a body with dry friction.

$\dot{x} = \mathrm{d}x/\mathrm{d}t$) determines the force direction, where x is the load displacement with respect to the base. The sum of the elastic force $P = -cx$ proportional to the displacement and the resistance force with the absolute value, R_0, defines the resultant force, F, i.e.,

$$F = -cx + R_0 \sin \dot{x} \qquad (1.2)$$

where c is the system stiffness.

Here the function $\sin \dot{x} = 1$ at $\dot{x} > 0$, $\sin \dot{x} = -1$ at $\dot{x} < 0$, $\dot{x} = 0$ at $\dot{x} = 0$. A closed hysteresis loop takes place in the cyclic motion of the load with the amplitude, A, on the plane (F,x) (see Figure 1.2). The hysteresis loop area equals the energy losses of the system in the vibration cycle

$$\Delta W = 4R_0 A$$

The amplitude value of the elastic potential energy of the system is determined as $W = cA^2/2$. The dissipation factor ψ defined from Equation (1.1) equals

$$\psi = 8R_0/cA \qquad (1.3)$$

Relative energy damping in the model with "dry" friction depends on the vibration amplitude, as follows from Equation (1.3).

Let us now consider *the model for a body with the frictional force proportional to the displacement*. This is the system shown in Figure 1.3, which consists of the load fastened on the plate spring. The plates of the spring are mounted without a preliminary tightness. The frictional force

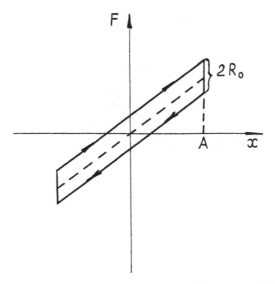

FIGURE 1.2. Hysteresis loop at harmonic vibration of the system with dry friction.

between the spring plates is proportional to the contact pressure which, in its turn, is proportional to the displacement. Figure 1.4 illustrates the plot of the force F that affects the load. The resultant force F is proportional to the displacement x during the system loading, viz.,

$$F = c_1 x \qquad (1.4)$$

where c_1 is the system stiffness during the increasing displacement. During

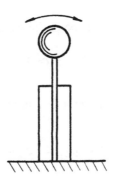

FIGURE 1.3. Model of a body with a resistance force proportional to the displacement.

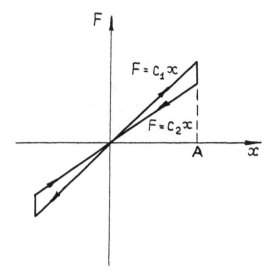

FIGURE 1.4. Hysteresis loop for the system with a resistance force proportional to the displacement.

unloading (decreasing the displacement absolute value) the stiffness coefficient equals c_2

$$F = c_2x \qquad (1.5)$$

and $c_0 = 1/2(c_1 + c_2)$ is the average stiffness. The area of the system hysteresis loop in harmonic vibrations is as follows

$$\Delta W = (c_1 - c_2)A^2$$

the amplitude value of the elastic potential energy, W, equals

$$W = \frac{1}{2} c_0 A^2$$

In this case the dissipation factor [Equation (1.1)] takes the following form

$$\psi = 2(c_1 - c_2)/c_0 \qquad (1.6)$$

The dissipation factor in this case does not depend on the amplitude and the frequency of the vibrations.

The resistance force for *bodies with viscous friction* is proportional to the velocity. The hysteresis loop in this case (see Figure 1.5) is elliptic in form. The load (a ball) on a spring in the viscous fluid is the mechanical model for the friction type under consideration (see Figure 1.6). The resistance force, R, from the fluid is proportional to the body velocity, $\dot{x}$

$$R = \alpha\dot{x} \tag{1.7}$$

where α is the viscosity factor. The resultant force, F, equals the sum of the resistance force [Equation (1.7)] and the elastic force:

$$F = cx + \alpha\dot{x} \tag{1.8}$$

where c is the stiffness coefficient.

In harmonic excitation of the system, in accordance with the law $x = A \cos \omega t$, the viscous friction force [Equation (1.7)] does the following work in one loading cycle

$$\Delta W = \int_0^{2\pi/\omega} R\dot{x}dt = \alpha A^2\omega^2\int_0^{2\pi/\omega} \sin^2 \omega t dt = \pi\alpha A^2\omega \tag{1.9}$$

The work equals the area of the hysteresis loop (see Figure 1.5). The elastic potential energy of the system has the amplitude value

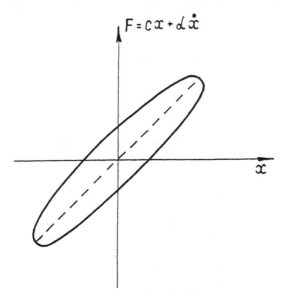

FIGURE 1.5. Hysteresis loop for a body with viscous friction.

$$W = \frac{1}{2} cA^2 \qquad (1.10)$$

The ratio of energy losses [Equation (1.9)] to the amplitude value of the elastic potential energy [Equation (1.10)] is the dissipation factor

$$\psi = 2\pi \frac{\alpha\omega}{c} \qquad (1.11)$$

which is directly proportional to the vibration frequency, ω, and does not depend on the vibration amplitude.

Figure 1.7 shows the main mechanical elements used in modeling the behavior of solids. Their combination makes it possible to build the more complex models shown in Figure 1.8. It is clear that the potential to build more complex combinations from the simplest elements represented in Figure 1.7 is practically unlimited. Modern computer methods have been developed to predict the model structure that describes best the behavior of the specific body. Reference [20], for example, gives a few examples of these methods.

1.2 THEORY OF LINEAR HEREDITARY ELASTICITY (VISCOELASTICITY)

The theory of viscoelasticity develops the simplest models of viscoelastic solids (Figure 1.6). It is based on more general integral relations between

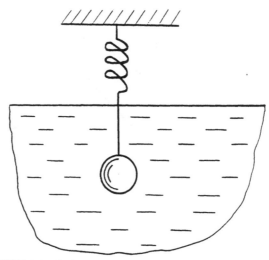

FIGURE 1.6. Mechanical model of a body with viscous friction.

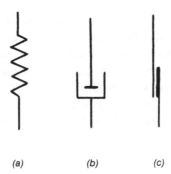

(a)   (b)   (c)

FIGURE 1.7. The simplest mechanical elements as the models for the behavior of the solids.

stresses and strains. The theory enables one to describe at one time both rheological processes in the static loading of solids (creep and stress relaxation) and the energy dissipation in vibrating solids under stationary and nonstationary excitation conditions.

Theories of linear viscoelasticity for isotropic and anisotropic bodies have been developed [19,65,73]. When considering energy dissipation for stationary monoharmonical loading of a viscoelastic body it is convenient to express the relation between stresses and strains in complex form and to introduce complex viscoelastic moduli or compliances. The general integral relationships between stresses and strains cover the simplest models of the viscoelastic body as well. A change from general integral relationships to

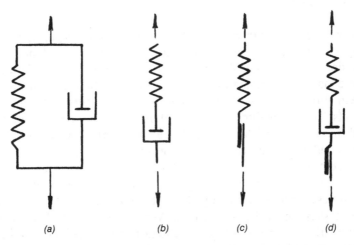

(a)   (b)   (c)   (d)

FIGURE 1.8. Main models: (a) Kelvin-Voigt model; (b) Maxwell model; (c) Prandtl model; and (d) Elastic-viscoplastic body.

the complex form makes it possible to effectively solve problems of stationary vibrations on the basis of the correspondence principle.

As the experiments show [46,61 – 64], the amplitude dependence of the relative energy dissipation for some materials lies within the elastic zone. The model of the linear viscoelastic body enables one to consider the frequency dependence of energy dissipation. However, it describes amplitude-independent internal friction in case the relative energy dissipation does not depend on the stress amplitude.

1.3 THEORY OF MICROPLASTIC DEFORMATION

The theory of microplastic deformation is a way to describe phenomenologically the behavior of nonuniform polycrystalline materials [56]. The theory rests on the assumption that loading of the material within elastic limits (when the macro volumes of the body exhibit elastic deformation) is followed by local plastic deformations in the micro volumes. The plastic deformations result in energy losses. In terms of the dislocation theory, this kind of energy dissipation relates to the motion of the dislocations. The Prandtl model [see Figure 1.8(c)] is the simplest model describing the phenomenon of microplasticity. The theory of microplastic deformations did not receive wide development.

1.4 C. M. ZENER'S THERMODYNAMIC THEORY OF VIBRATION DAMPING

The thermodynamic theory of C. M. Zener [94] holds a peculiar position among the theories of energy dissipation. According to Zener, energy dissipation is an inalienable property of any ideally elastic body. It is believed that energy dissipation results from irreversible heat fluxes arising between different parts of the vibrating body because of temperature gradients. The fluxes (both of micro and macro nature) result from the fact that different parts of the material are under stress states of different intensity and hence have different temperatures.

For example, in a bar under bending, one region of it appears to be compressed, while another one appears to be tensile. The temperature of the compressed region is slightly above the temperature of the tensile one (the process is not ideally isothermal). As a result of thermal conductivity, heat flux arises from the compressed region to the tensile one. The irreversible nature of the heat flux results in the dissipation of the mechanical vibration energy and provides its transfer to heat energy.

By Zener's theory, the following formula defines the coefficient of energy damping:

$$\psi = \frac{\Delta W}{W} = 2\pi \frac{E_{ad} - E_{is}}{E_{ad}} \frac{\omega_0 \omega}{\omega_0^2 + \omega^2} \qquad (1.12)$$

Here ΔW and W are the energy losses in a vibration cycle and the amplitude energy value respectively, E_{ad} and E_{is} are the adiabatic and the isothermal material moduli of elasticity, respectively, ω is the frequency of free vibrations, and ω_0 is the so-called relaxation frequency that is determined from the formula:

$$\omega_0 = \frac{\pi \lambda}{2 c_p \varrho h^2} \qquad (1.13)$$

where λ is the coefficient of heat conductivity, c_p is the heat capacity, ϱ is the material density, h is the thickness of the vibrating sample.

Figure 1.9 illustrates Equation (1.12). The parameter of energy dissipation has the maximum value peculiar to the theory, when the frequency equals the relaxation frequency, ω_0.

The example discussed above falls in the case of a body (a sample) under

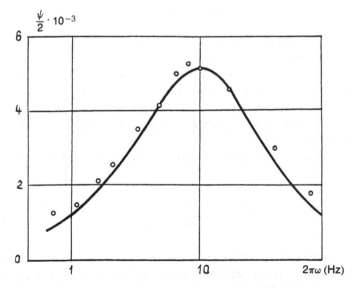

FIGURE 1.9. Diagram of energy dissipation by C. M. Zener: —, theoretical curve, O, experimental points.

nonuniform stress state (bending). Real solids are also known to dissipate the energy under uniform (on the average) stress state (for example, under uniaxial tension or compression). For this case Zener pointed to one more mechanism of damping. This is the microstructural mechanism common for polycrystalline bodies. The symmetry axes of individual crystals are randomly arranged in such bodies. Therefore, local temperature differences will arise at the equal average strain value. Consequently, local heat fluxes appear. Energy damping associated with this effect reaches a maximum at the relaxation frequency that is described by the formula:

$$\omega_0 = \frac{3\pi\lambda}{\varrho\, c_p a^2} \tag{1.14}$$

where a is the average linear size of the crystal in the polycrystalline body.

The interior elegance of Zener's theory is highly effective. However, it follows from the theory that there is evident dependence of dissipation parameters on sample linear dimensions [see Equation (1.13)] as well as a difference of dissipation parameters in bending and tension-compression. These facts are in rather poor agreement with practical experience and common sense. On the whole it is believed that Zener's theory is of limited utility for describing energy dissipation in real materials.

1.5 THEORY OF ELASTIC HYSTERESIS

The imperfect elasticity of materials shows itself at stresses below the elastic limit, and lies in the fact that the relationship between stresses and strains is nonlinear and ambiguous under loading and unloading. It results in the formation of a hysteresis loop on the stress-strain curve. Direct description of the loop is a way to describe energy dissipation effects in the materials. Such a method of approach was developed by the Ukrainian scientific school [33,46,61,62,93].

For example, the form of the hysteresis loop in a uniaxial stress state is as follows

$$\overset{\leftrightarrow}{\sigma} = E\left[\epsilon \pm \frac{3}{8}\delta\left(\epsilon_0 \mp \frac{\epsilon^2}{\epsilon_0}\right)\right] \tag{1.15}$$

where E is the material's modulus of elasticity, δ is the logarithmic decrement of vibrations, ϵ_0 is the amplitude strain value, the arrows over the stress, σ, in Equation (1.15) correspond to material loading and unloading.

The area of the hysteresis loop characterizes the value of the energy damped in a material unit volume and depends on the stress or strain amplitude. The decrement of vibrations in Equation (1.15) may be considered as a function of the load frequency, temperature, and other factors.

Solution of the problems on the basis of the given theory demands rather labor-consuming methods of nonlinear mechanics, e.g., derivation of slightly nonlinear differential equations in the form of asymptotic expansion with respect to the order of the small parameter [61].

The theory of elastic hysteresis enables one to solve the problems of stationary vibrations with consideration of energy dissipation in the materials. However, the theory was not generalized for the case of complex stress states, anisotropic materials, and transition vibration regimes.

1.6 ENERGY METHOD FOR CONSIDERATION OF THE INTERNAL FRICTION

The approximate energy method is widely used [38,48,57,69]. It is believed that the linear stress-strain relationship holds. Additional hypotheses about the energy dissipation function in a loading cycle are introduced, so that the dissipative parameters of a material are considered to be independent along with material elastic constants. For example, logarithmic decrements of vibrations or dissipation factors are such parameters.

Application of the energy method usually involves determining system vibration frequencies and modes (the system is believed to be perfectly elastic) and the subsequent use of energy balance equations for approximate prediction of vibration amplitudes. The energy method is the most general one in a sense, as it deals directly with the damped energy and does not relate it to any specific internal mechanism (viscoelasticity, microplasticity, etc.). This method enables one to consider phenomenologically the effect of different factors on the energy dissipation.

Use of the energy method for describing energy dissipation in anisotropic bodies under an arbitrary stress state is presented in detail in the following chapters.

CHAPTER 2

Methods for Experimental Investigation of Energy Dissipation Response

There are two groups of methods used to examine the characteristics of energy dissipation in the materials under cyclic loading. The methods of the first group are the direct methods, based on the direct measurements of dissipated energy values. The methods of the second group are indirect methods, where the amount of dissipated energy is determined from changes in the other parameters (such as amplitudes, frequencies, lags, etc.). Energy and thermal methods, as well as the method of the hysteresis loop, fall into the group of direct methods. The method of free damped vibrations, the method of the resonance curve, and the phase method fall in the group of indirect methods.

2.1 DIRECT METHODS FOR DETERMINING ENERGY LOSSES

2.1.1 The Energy Method

The energy method is based on direct measuring the electrical or mechanical excitation power needed to maintain steady-state vibrations in the sample under investigation. The complete power N can be divided into two parts. Let us denote them as N_1 and N_2. N_1 is the power that is expended for energy dissipation in the material during vibrations. N_2 is the second part of the complete energy, which is spent to overcome the resistance in the excitation system, aerodynamic losses, etc. To separate the power N_1, it is common to perform tests on specimens made of the materials with negligible internal friction, as compared to the sample under investigation. The power measured in these tests is believed to be the power N_2.

The relative energy dissipation in the material under vibrations is determined by the formula

$$\psi = \frac{N - N_2}{\omega W} \tag{2.1}$$

where ω is the frequency of the steady-state vibrations, and W is the potential elastic energy of the specimen for the corresponding amplitude. The method is not widely used because of the error involved in determining the power N_2.

2.1.2 The Thermal Method

The thermal method to determine energy dissipation characteristics is based on the hypothesis that the majority of energy lost due to internal friction is transformed to heat energy, with the result that the specimen is heated. The extent of heating depends on the quantity of heat liberated per unit of time (the power of dissipation), i.e., it depends on the vibration frequency. The heat energy, Q, can be measured by the calorimetric method. The method involves measuring the temperature of the working substance cooling the sample under stable cyclic deformation with the frequency, ω, in a definite time, t. The temperature difference of the working substance at entry and exit of the calorimeter, ΔT, and the working substance mass, m, in a time, t, define the heat energy as follows

$$Q = m\Delta T \tag{2.2}$$

The dissipated energy value in a complete loading cycle is calculated in the following way

$$\Delta W = Q/t\omega \tag{2.3}$$

The energy dissipated in real vibrating bodies transforms not only into heat energy, but is also spent for irreversible changes in the internal structure of the material (crack growth, dislocation movement, etc.). This fact and the difficulties in technical realization constrain the application of the thermal method.

2.1.3 The Method of the Hysteresis Loop

The method of the hysteresis loop includes direct definition of a dynamic or static hysteresis loop of the stress-strain curve ($\sigma \sim \epsilon$) of the specimen under the cyclic loading (see Figure 2.1). The area of the hysteresis loop determines energy losses, ΔW, in the material in a loading cycle, which

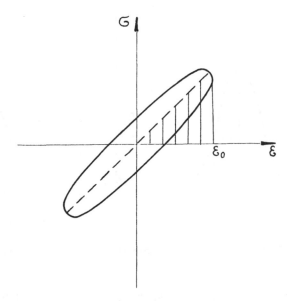

FIGURE 2.1. Hysteresis loop in the stress-strain curve ($\sigma \sim \epsilon$) of a body under cyclic loading.

correspond to the strain amplitude, ϵ_0. The area of the shaded triangle in Figure 2.1 equals the amplitude of the potential elastic energy, W.

As a rule, the relative energy dissipation ($\psi = \Delta W/W$) for real materials is given in units of percent or decimal fractions of a percent. That is why the area of the hysteresis loop is very small. It requires, in its turn, high accuracy in strain measurement during the stress-strain curve registration.

2.2 INDIRECT METHODS FOR DETERMINING ENERGY LOSSES

2.2.1 The Method of the Resonance Curve

Forced vibrations of the sample are set up by an external harmonic force with a constant amplitude Q_0 (the force excitation) or by a harmonic displacement with a constant amplitude A_0 (the kinematic excitation). Since the excitation frequency, ω, is variable, it is possible to obtain a curve with the resonance peak of the displacement [the force excitation, Figure 2.2(a)] or a curve with the resonance hollow of the force [the kinematic excitation, Figure 2.2(b)]. Damping characteristics are calculated by measuring the

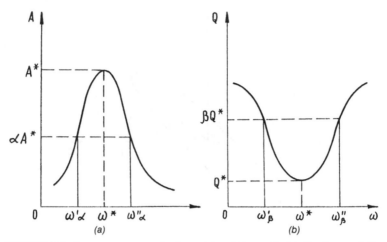

FIGURE 2.2. Resonance curves for force (a) and kinematic (b) vibration excitations.

resonance peak ($\omega_\alpha'' - \omega_\alpha'$) or resonance hollow ($\omega_\beta'' - \omega_\beta'$) width at the level of the values αA^* (βQ^*), where A^* (Q^*) is the resonance amplitude of the displacement (the force). It is customary to assume that the coefficients α (β) are equal to 0.5 or 0.7. The methods mentioned make it possible to obtain the amplitude dependence of the damping factor from one resonance curve.

2.2.2 The Method of Free Damped Vibrations

The method of free damped vibrations has gained wide acceptance because it is vivid and simple in technical realization (excitation of vibrations). Once the sample is disturbed from an equilibrium condition, it is left on its own, i.e., it executes free vibrations with an energy dissipation; this registers on the vibrogram (see Figure 2.3). Based on the vibrogram, the values of damping characteristics are calculated.

The rate of free vibrations reduction is usually estimated by the logarithmic decrement of vibration, δ, or by corresponding relative energy dissipation, i.e., the dissipation factor, ψ. The logarithmic decrement, δ, can be determined in one cycle of the vibrations, i.e., in a period of the vibrations, T. It also can be determined in n cycles, of the vibrations from the values of displacement amplitudes with the help of the vibrogram (Figure 2.3)

$$\delta = \frac{1}{n} \ln \frac{A_i}{A_{i+n}} \qquad (2.4)$$

where A_i, A_{i+n} are the amplitudes corresponding to the ith and the $(i + 1)$th cycles of the vibrations. If $n = 1$, then

$$\delta = \ln \frac{A_i}{A_{i+1}} \qquad (2.5)$$

The decrements [Equations (2.4) and (2.5)] can depend on the value of the displacement amplitude. Because of this, the problem of amplitudes selection is of prime methodological importance. It is advisable to take the mean amplitude, A_m, as the amplitude of the vibrations

$$A_m = (A_i + A_{i+n})/2 \qquad (2.6)$$

The dissipation factor, ψ, corresponding to n cycles of the vibrations and the mean amplitude, A_m [Equation (2.6)], is defined as the ratio of mean energy losses in n cycles, ΔW, to the mean level of the system elastic potential energy, W_m, i.e.,

$$\psi = \Delta W / W_m \qquad (2.7)$$

Let us find the relation between Equations (2.4) and (2.7), using the equation of the energy balance for the time period, nT. Let us consider the system states at the moments when the displacement amplitude reaches the

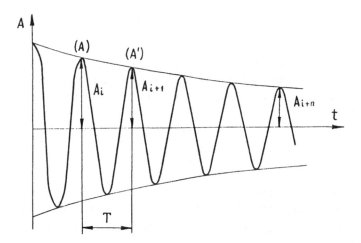

FIGURE 2.3. A vibrogram of free damped vibrations.

values A_i and A_{i+n}. The kinetic energy of the system equals zero at these states. For the simplest systems the level of the potential elastic energy is determined from the amplitude value of the displacement and the system stiffness c. The equation of the energy balance is of the form

$$cA_i^2/2 = cA_{i+n}^2/2 + \Delta W \tag{2.8}$$

In Equation (2.7) the potential energy, W_m, that corresponds to the mean amplitude of the displacements in n cycles [Equation (2.3)] is expressed as follows

$$W_m = c(A_i + A_{i+n})^2/8 \tag{2.9}$$

Considering Equations (2.7), (2.8) and (2.9), one will obtain

$$\psi = \frac{4(A_i - A_{i+n})}{n(A_i + A_{i+n})} \tag{2.10}$$

or for one cycle of the vibrations ($n = 1$)

$$\psi = \frac{4(A_i - A_{i+1})}{(A_i + A_{i+1})} \tag{2.11}$$

Let us expand $\ln (A_i/A_{i+n})$ [Equation (2.5)] into a series, i.e.,

$$\delta = \sum_{j=0}^{\infty} \frac{2}{2j + 1} \left(\frac{A_i - A_{i+n}}{A_i + A_{i+n}} \right)^{2j+1}$$

It can be shown (see Reference [62], for example) that the value of the logarithmic decrement, δ, is determined by the first term of the series with sufficient accuracy. That is why Equation (2.5) can be exchanged for the approximate one

$$\delta = \frac{2(A_i - A_{i+n})}{n(A_i + A_{i+n})} \tag{2.12}$$

Comparing Equations (2.10) and (2.12), it is possible to arrive at an approximate agreement between the logarithmic decrement of the vibrations [Equation (2.4)] and the dissipation factor [Equation (2.7)]. It is as follows

$$\psi \approx 2\delta \tag{2.13}$$

    The values of the logarithmic decrement [Equation (2.4)] and the dissipation factor [Equation (2.7)] are the averaged characteristics of the energy dissipation in n cycles of the vibrations. Their values, calculated from Equations (2.5) and (2.11), are believed to be true or nonaveraged, which correspond to each cycle of the vibrations with the average amplitude of the displacement $A_m = (A_i + A_{i+1})/2$. For amplitude-independent internal friction, Equations (2.4) and (2.5) coincide with one another.

    There are two main methods that are used in practice to determine the damping characteristics of different materials. These are the method of free damped vibrations and the method of the resonance curve. Technical realization of the methods differs in the techniques of the specimen loading. The following techniques can be indicated: longitudinal (a) and torsional (b) vibrations, transverse vibrations of cantilever specimens (c), vibrations under ideal bending (d), and transverse vibrations of the free supported specimen (e). Figure 2.4 illustrates the respective schemes.

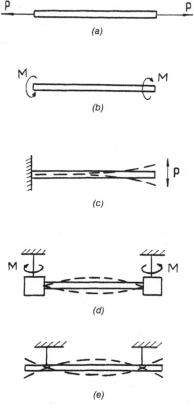

FIGURE 2.4. The schemes of specimen loading.

When the specimen is under ideal bending [Figure 2.4(d)], two massive loads are applied to its ends. The loads are able to rotate about parallel axes. In all cases of loading, both free damped vibrations and forced vibrations can be realized. As a rule, the schemes (a), (b), and (c) are used in the case of forced vibrations and the schemes (c) and (d) are used in the case of free damped vibrations.

All schemes of loading, except for the first one (a), conform to the complex stress state of the specimen. That is why in order to obtain the energy dissipation characteristics of the material in terms of the energy dissipation characteristics of the sample, it is necessary to use additional hypotheses of energy dissipation under corresponding loading conditions.

Excitation of free damped vibrations (in all loading cases mentioned above) takes place due to the removal of the links of the preliminary static loading. The specimens vibrate in accordance with the first mode that has the least power of energy dissipation as compared to the next vibration modes.

The forced vibrations are excited, as a rule, with the help of a noncontact electromagnetic disturbance. This makes it possible to change the force and the frequency of excitation and to get different vibration modes and amplitude-frequency characteristics of the specimens.

It should be noted that dynamic energy losses for the bearing friction, aerodynamic friction, etc. require additional estimation for each of the methods.

Of course, the authors do not aspire to completely describe the large and continually expanding list of methods for measuring and the schemes for representing specimen loading when determining the damping behavior of the materials. Specific schemes of test devices can be found elsewhere [25,46,62,93].

2.3 CONSIDERATION OF THE AERODYNAMIC COMPONENT OF ENERGY DISSIPATION

The experiments to determine damping behavior are often performed under standard atmospheric pressure. Aerodynamic resistance forces affect any moving (vibrating) body. That is why, in parallel with the internal friction of a vibrating material, one should also take into consideration the work done by aerodynamic forces.

A set of experiments to determine the damping response of fibrous composites points to the necessity for air medium effect consideration [3,25,26,47]. The specimens were tested in vacuum and at atmospheric pressure. Experimental data obtained enable one to estimate the effect of the energy dissipation aerodynamic component on the dissipation factor value. If the specimen has a low internal friction and low stiffness, then the dissipation factor under bending due to aerodynamic losses can be many times higher than the true one [3,25].

It is common practice to determine the material damping behavior under specimen bending vibrations. Let us consider the work done by aerodynamic forces in vibrating plane beams for two widespread types of specimen fastening — cantilever and free double-seat types (see Figure 2.5).

Let the aerodynamic force acting on each specimen element be proportional to the standard air density, ϱ_a, and the square of the element velocity, v [vi:] (see Figure 2.6). Let us believe that in the case of low specimen vibrations the unit aerodynamic force vector holds its vertical direction. The aerodynamic force dN directed along the z-axis affects the specimen element of width dx

$$dN = c_n \frac{\varrho_a v^2}{2} b dx \tag{2.14}$$

where b is the specimen width, and c_n is the proportionality constant.

Aerodynamic resistance forces do the work $\Delta\Pi_a$ in the vibration cycle for the specimen as a whole

$$\Delta\Pi_a = 4 \int_0^l \int_0^{Z_m(x)} b c_n \frac{\varrho_a v^2}{2} dx \tag{2.15}$$

Here $Z_m(x)$ is the maximum deflection (the vibration amplitude) in the section with the coordinate x, v [vi:] is the velocity of the section, and l is the specimen length. The factor 4 is before the integral in Equation (2.15) because the integral defines the work in a quarter of the vibration period. Let us change the velocity, v [vi:], in Equation (2.15) for the derivative of

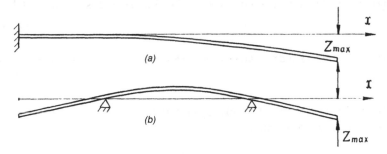

FIGURE 2.5. Schemes of fastening for plane samples: (a) cantilever and (b) free double-seat.

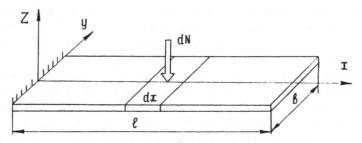

FIGURE 2.6. Geometrical parameters of the cantilever sample.

the coordinate $dz/dt = \dot{z}$, and then turn to integration over the time, t, in the period quarter, $T/4$. Then one will obtain

$$\Delta\Pi_a = 2bc_n\varrho_a \int_0^l \int_0^{T/4} \dot{z}^3 \, dt \, dz \tag{2.16}$$

Let us consider specimen vibrations in the first natural mode for two cases:

(1) Vibrations of the cantilever plane specimen
(2) Vibrations of the free double-seat plane specimen (see Figure 2.5)

The solution to the equation of beam bending vibrations is of the form [82]

$$Z(t,x) = \sin \omega t \, Z_{max} \, X(x)/X(l) \tag{2.17}$$

where Z_{max} is the maximum sample deflection, ω is the circular vibration frequency, and $X(x)$ is the normal mode of vibrations. The following functions correspond to the first vibration modes in the cases under consideration:

$$X(x) = -0.734 \sin ax + \cos ax + 0.734 \sinh ax - \cosh ax, al = 1.875 \tag{2.18a}$$

$$X(x) = -0.983 \sin ax + \cos ax - 0.983 \sinh ax + \cosh ax, al = 4.730 \tag{2.18b}$$

Here al are the roots of frequency equations, a is the parameter defined from the following formula

$$a^2 = \omega \sqrt{\frac{\varrho_m \, bh}{D}} \tag{2.19}$$

Here ϱ_m is the density of the specimen material, h is the specimen thickness, and D is the specimen flexural stiffness

$$D = E_x bh^3/12$$

E_x is the modulus of elasticity of the material along the x direction.

Now let us compare the relative energy damping (the dissipation factors) due to the internal friction and to aerodynamic losses. Let us relate the work of aerodynamic drag forces [Equation (2.16)] to the amplitude value of the potential elastic energy, Π, of the samples. The value of Π can be expressed from the amplitude of the bending moment, M

$$\Pi = \int_0^l \frac{M^2}{2D}\, dx \tag{2.20}$$

where

$$M = D\frac{\partial^2 z}{\partial x^2}\Bigg|_{\sin \omega t\,=\,1} \tag{2.21}$$

Using Equations (2.17) and (2.21), the expression for the amplitude of the potential energy, Π [Equation (2.20)], takes the form

$$\Pi = \frac{DZ_{max}^2}{2X^2(l)} \int_0^l \left(\frac{\partial^2 X}{\partial x^2}\right)^2 dx \tag{2.22}$$

Let us take the ratio of aerodynamic force work in a vibration cycle [Equation (2.16)] to the amplitude value of the elastic potential energy [Equation (2.22)] as the coefficient of aerodynamic losses, ψ_a. Using Equations (2.16), (2.17), and (2.19), one will obtain the following formula:

$$\psi_a = c_n\beta\,\frac{\varrho_a}{\varrho_m}\frac{Z_{max}}{h} \tag{2.23}$$

where coefficient β depends on the sample vibration mode [Equations (2.18)]

$$\beta = \frac{8a^4 \int_0^l X^3\, dx}{3X(l) \int_0^l \left(\frac{\partial^2 X}{\partial x^2}\right)^2 dx} \tag{2.24}$$

Equations (2.18) make it possible to calculate the value of β for the first vibration modes of the sample for two types of fastening under consideration:

(1) $\beta = 1.98$ (cantilever)
(2) $\beta = 0.57$ (free double-seat)

It follows from Equation (2.23) that the coefficient of aerodynamic losses is proportional to the ratio between the medium and sample densities, and to the relative amplitude of the displacements.

Assume that energy damping results only from internal friction and aerodynamic losses. Then to obtain the true material dissipation factor, it is sufficient to determine the difference between the dissipation factor calculated from the vibrogram of free damped vibrations and the coefficient of aerodynamic losses which corresponds to the amplitude of sample vibrations.

If needed, this procedure can also be used for the higher vibration modes, for different types of sample fastening. In doing so one must take into account the changes in Equations (2.18) and (2.24). The formulas obtained have an approximate character, i.e., they are adequate only for approximate calculations.

Let us note that the works [11,26,27] give the formula for relative energy damping due to aerodynamic losses for the cantilever sample. This formula

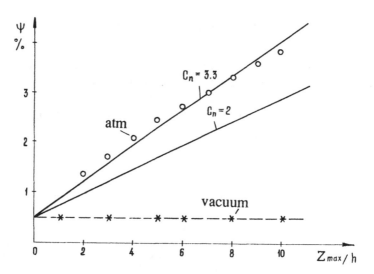

FIGURE 2.7. Dissipation factors of aluminum cantilever samples versus the ratio between maximum amplitude of deflection, Z_{max}, and the sample width, h. Solid lines are for theoretical calculations, points are for experimental data [26]: (O) atmosphere pressure, (*) vacuum.

is similar to Equation (2.23), if one changes the product βc_n to the coefficient k. The works do not give the value of coefficient k.

The authors of Reference [26] investigated aluminum cantilever samples in tests at atmospheric pressure and in vacuum. They presented experimental relationships between relative energy damping and the amplitude of free damped vibrations (see Figure 2.7). As is seen, the dissipation factor value in vacuum, ψ_0, practically does not depend on the amplitude of vibrations. Therefore, the factor, ψ, at atmospheric pressure can be considered as the sum of the constant component, ψ_0, and the coefficient of aerodynamic losses, ψ_a. The authors of the book calculated the values of ψ_a. The following values were used in calculations according to Equation (2.23): $\varrho_a = 1.293$ kg/m³, $\varrho_m = 2700$ kg/m³, $\beta = 1.98$.

Experimental data from References [3], [11], [25], [26], and [47] verify the linear character of the relationship between the coefficient of aerodynamic losses [Equation (2.23)] and the amplitude of vibrations.

CHAPTER 3

General Information on Energy Dissipation in Composites

The literature on the experimental investigation of energy dissipation in composite materials is quite extensive, which is not surprising. The variety of fiber and matrix types, reinforced structures and loading types, as well as the variety of environmental conditions wherein the materials are applied is so wide that the process of accumulation of experimental information on composite dissipation behavior is far from completion.

The present chapter includes data on experimental examination of composite dissipative properties which have been published in the world literature and obtained by the authors. There are two ways to use this information. The first one is to directly apply it if the material type and test conditions coincide with practical requirements. The second way is to use the experimental information to develop a theoretical model for a phenomenon. This second way enables one to use the available experiments for prediction of the properties of new materials in a wide range of loading conditions. Unfortunately, a great number of published experimental works contain only fragmentary and incomplete information on the materials under investigation. This makes the use of the information difficult for theoretical generalizations or for construction of theoretical models.

Experimental data presented in the chapter are partitioned into three groups. The first group of data shows the effect of the fiber relative fraction and the fiber length on the dissipative response of unidirectional fibrous composites. It also includes the information on the anisotropy (the dependence of the properties on the load direction) of the dissipative behavior of unidirectional fibrous composites. The second group gives the information on dissipative properties of multilayered fibrous composites with different orientation of the layers under uniaxial loading and bending. The data of the third group show the effect of the load frequency, the load amplitude, temperature, etc., on composite dissipative parameters.

Certainly, the information presented in this chapter is not an exhaustive review of experimental works. The references at the end of the book will form a more complete impression.

3.1 DISSIPATIVE BEHAVIOR OF UNIDIRECTIONAL COMPOSITES

Experimental investigations of unidirectional glass and carbon fiber reinforced plastics (GFRP and CFRP) were carried out in Reference [92]. The specimens of GFRP and CFRP were tested by the method of the resonance curve under bending vibrations on two supports. The fibers were directed along the specimen axis. Primary attention was paid to the effect of the fiber volume fraction, the temperature, and the load frequency on elasto-dissipative properties of the composites. The "loss factor" (loss tangent, η) was selected as the damping characteristic. The loss tangent relates to the relative energy dissipation, ψ, as follows: $\psi = 2\pi\eta$. The plots of Young's modulus and the loss factor as the functions of the fiber volume fraction in polyester composites, ξ, at 25°C are given in Chapter 5 where they are compared with theoretical curves. There are experimental data for the composites with the other fiber and matrix types in the work. The authors showed that the temperature effect on composite dissipative parameters depended heavily on the matrix type; the loss factor varied slightly when the load frequency varied from 70 to 700 Hz.

The effect of the fiber length and the load frequency for different values of the fiber fraction in GFRP was investigated in Reference [58]. The plots in Figure 3.1 show Young's modulus, E, and the loss factor, η, versus the load frequency for three values of the fiber fraction: 0, 10, and 20%. Figure 3.2 illustrates the effect of the fiber length on composite elasto-dissipative parameters along the fibers and in the transverse direction. The diameter of the fibers was 10 μm, the test frequency was 300 Hz.

Epoxy composites on the base of graphite, aramid, and boron fibers were studied in Reference [78]. Figures 3.3 and 3.4 show the effect of the relative fiber length (the ratio of the fiber length to its diameter) on elasto-dissipative properties of CFRP.

It might be well to point out that the effect of the fiber relative fraction on the composite damping response was also examined in References [1] and [4].

The energy dissipation anisotropy for a set of fibrous composites was examined in Reference [2]. The experiments were carried out for unidirectional, angle-plied specimens and the specimens with more complex structures. The specimens were tested under bending and torsion. The experimental data for unidirectional composites (the monolayer properties)

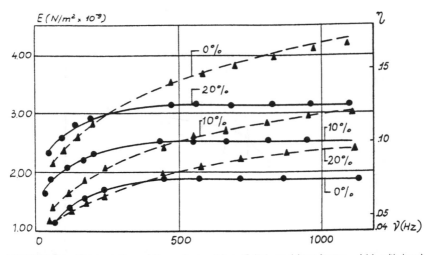

FIGURE 3.1. The variation of Young's modulus, E (●), and loss factor η (▲), with load frequency in unidirectional polypropylene GFRP. Fiber concentration by weight is indicated in percent.

are included in Table 3.1. The rest of the experimental data from the work are given in Chapter 8 in comparison with predicted diagrams.

Reference [52] develops the previous work. Extensive experimental results along with theoretical diagrams are presented for epoxy CFRP and GFRP (HMS/DX 210 and GLASS/DX 210). Elasto-dissipative characteristics of unidirectional materials are included in Table 3.1. Experimental

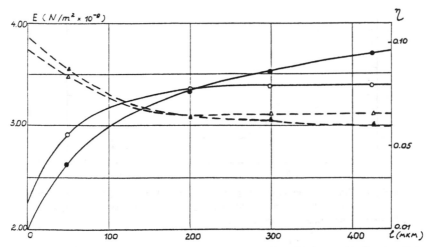

FIGURE 3.2. The variation of Young's modulus, E, and loss factor, η, with fiber length in unidirectional polypropylene GFRP. E (●) and η (▲) along the fibers; E (○) and η (△) in the transverse direction.

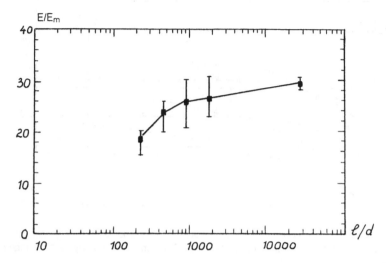

FIGURE 3.3. Plot of experimental relative Young's modulus along the fibers, E/E_m, as a function of fiber aspect ratio for unidirectional CFRP. Here E_m is the matrix modulus of elasticity. Fiber volume fraction equals 0.654. The points show average values and the segments show experimental scatter.

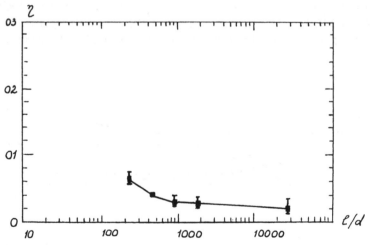

FIGURE 3.4. Plot of experimental loss factor, η, as a function of fiber aspect ratio for unidirectional CFRP. Fiber volume fraction equals 0.654. The points show average values and the segments show experimental scatter.

30

TABLE 3.1. *Elasto-Dissipative Properties of Unidirectional Epoxy Composites.*

Material	E_1 (GPa)	E_2 (GPa)	G_{12} (GPa)	ν_{12}	ψ_1^* (%)	ψ_2^* (%)	ψ_6^* (%)	References
HMS/DX 209	188.8	6.01	2.7	0.3	0.64	6.9	10.0	[2]
HTS/DX 210	103.4	7.6	3.8	0.3	0.49	5.48	6.75	[2]
HMS/DX 210	172.7	7.2	3.8	0.3	0.45	4.22	7.05	[52]
GLASS/DX 210	37.8	10.9	4.9	0.3	0.87	5.05	6.91	[52]
KMU-3	139.0	9.0	5.2	0.17	4.7	7.4	11.0	Authors
KMU-4	100.4	10.1	6.9	0.21	3.5	6.4	10.0	Authors
"Kulon"	238.0	4.4	3.7	0.2	3.6	10.1	16.15	Authors

Here E_1 is the Young's modulus along the fibers; E_2 is the Young's modulus in the transverse direction; G_{12} is the shear modulus in the transverse direction; ν_{12} is the major Poisson's ratio; and ψ_1^*, ψ_2^*, and ψ_6^* are the dissipation factors under uniaxial loading along the fibers, in the transverse direction, and under ideal shear in the monolayer plane, respectively.

data for Young's modulus, E, and the dissipation factor, ψ, of unidirectional specimens under bending versus the angle between the fibers and the specimen axis are given in Chapter 4 as compared to predicted results.

Reference [54] deals with unidirectional hybrid glass-carbon fiber reinforced plastics. The materials consist of monolayers made of HMS/DX 210 and GLASS/DX 210 with different ratios of the monolayers. Experimental data from this work are used in Chapter 8.

The authors of this book performed experimental investigation of the dissipative behavior of several fibrous composites. Plane cantilever specimens were tested by the method of free damped vibrations. In particular, the authors studied the effects of aerodynamic damping (the experiments were realized in atmospheric conditions) and specimen fastening on elastic and dissipative parameters. Figures 3.5–3.7 present the dissipation factor, ψ, and Young's modulus, E, versus the angle between the fiber direction and the specimen axis for three types of CFRP. The authors' experimental data are also presented in Table 3.1.

3.2 DISSIPATIVE PROPERTIES OF MULTILAYERED FIBROUS COMPOSITES

The results of the experiments on the energy dissipation in vibrating multilayered beams and plates are present in some works mentioned previously. The effect of the angle between the bending moment plane and the

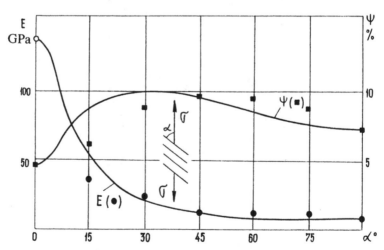

FIGURE 3.5. Dissipation factor, ψ (■), and Young's modulus, E (●), versus an angle between fiber direction and load direction, α, for unidirectional CFRP KMU-3 (Russia). Solid lines are for theoretical diagrams (see Chapter 4).

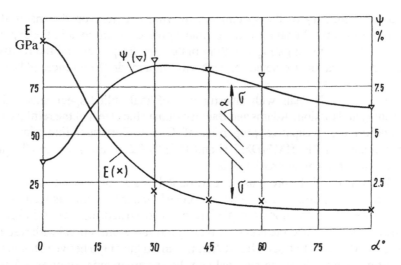

FIGURE 3.6. Dissipation factor, ψ ($\triangledown$), and Young's modulus, E ($\times$), versus an angle between fiber direction and load direction, α, for unidirectional CFRP KMU-4 (Russia). Solid lines are for theoretical diagrams (see Chapter 4).

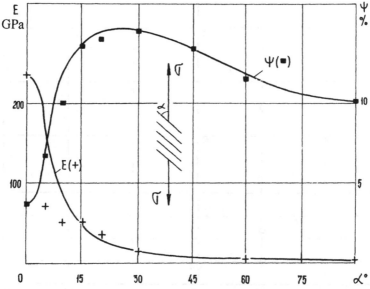

FIGURE 3.7. Dissipation factor, ψ ($\blacksquare$), and Young's modulus, E (+), versus an angle between fiber direction and load direction, α, for unidirectional CFRP "Kulon" (Russia). Solid lines are for theoretical diagrams (see Chapter 4).

axis of symmetry of the specimen's internal structure on the material's elasto-dissipative behavior is the primary problem examined in the experiments. When investigating vibrating plates, the frequency and the relative energy dissipation are noted for each vibration mode (References [44] and [54]).

Reference [2] deals with investigation of angle-plied specimens under bending and torsion. Adams and Bacon estimate the effect of the reinforcing angle on the dissipative properties of CFRP. Corresponding plots for angle-plied CFRP (HMS/DX 209 and HTS/DX 210) are given in Chapter 8 in comparison with predicted data.

Reference [52] presents experimental data on elasto-dissipative parameters for cross-plied CFRP and GFRP under bending, as well as for the materials with more complex reinforced structures. All reinforced structures of the specimens under investigation are symmetrical with respect to the mid-plane. In tests, Ni and Adams varied the angle between the main reinforcement axis of the material and the specimen axis. Figures 3.8 and 3.9 illustrate experimental results, but the main part of the results will be given in Chapter 8 as compared with predicted curves.

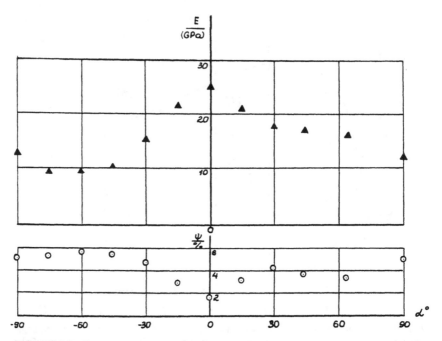

FIGURE 3.8. Flexural modulus, E (▲), and dissipation factor, ψ (⊙), versus outer layer fiber orientation angle for epoxy GFRP (GLASS/DX 210) specimens. Specimen structure: $(0°, -60°, 60°)_s$.

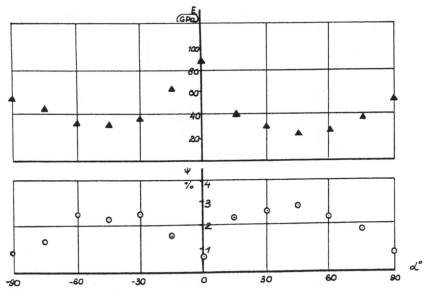

FIGURE 3.9. Flexural modulus, E (▲), and relative energy dissipation, ψ (⊙), versus outer layer fiber orientation angle for epoxy GFRP (HMS/DX 210) specimens. Specimen structure: $(0°,90°,45°,-45°)_s$.

The authors carried out their own tests to determine the elasto-dissipative behavior of CFRP under bending. The authors investigated CFRP KMU-4 and "Kulon." Figures 3.10 and 3.11 show the experimental results. All specimens had symmetrical structures with respect to the mid-plane. The specimens of KMU-4 had eight monolayers and the ones of "Kulon" had four monolayers.

3.3 EFFECT OF LOAD CONDITIONS ON COMPOSITE DISSIPATIVE RESPONSE

Some works concern experimental data on the composite elasto-dissipative behavior depending on load frequency, temperature, stress amplitude, and other factors. The most extensive information on the composite elasto-dissipative behavior with consideration of the effect of the load frequency, temperature, and the specimen vibration mode is given in Reference [25]. Here, Georgi estimated the effect of the stress amplitude on composite damping properties, took into account the aerodynamic damping, and determined the effect of atmospheric pressure on the experimental values of logarithmic decrements of vibrations. The experiments were realized not only with the use of double-seat and cantilever beams; composite structure

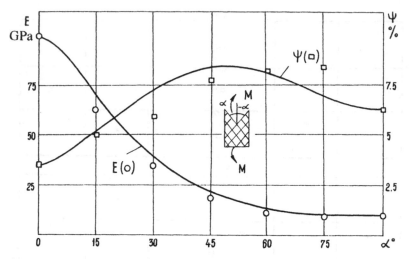

FIGURE 3.10. Dissipation factor, ψ ($\square$), and Young's modulus, E ($\bigcirc$), versus orientation angle, α, for angle-plied CFRP KMU-4. Solid lines are for theoretical diagrams (see Chapter 8).

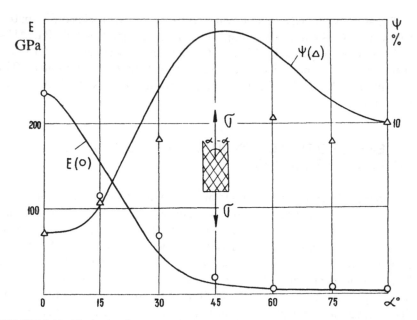

FIGURE 3.11. Dissipation factor, ψ ($\triangle$), and Young's modulus, E ($\bigcirc$), versus orientation angle, α, for CFRP "Kulon." Solid lines are for theoretical diagrams (see Chapter 8).

elements were also used, such as panels and helicopter propellers. Double-seat beams were used in free vibration tests with moderate stress amplitudes; six transverse vibration modes and one longitudinal vibration mode of the specimens were considered. Figures 3.12−3.23 present the experimental results for boron fiber reinforced plastic (BFRP) and CFRP. Elasto-dissipative parameters are shown as the functions of vibration modes, the fiber orientation, and the temperature. The specimens had reinforcing angles from the following set: 0°, ±22.5°, ±45°, ±67.5°, and 90°; the specimen structures were symmetrical with respect to the mid-plane. BFRP and CFRP consisted of eight and four layers, respectively.

Reference [26] covers experimental data on the dependence of the loss factor η upon the strain amplitude and the load frequency for cross-plied GFRP. The tests were carried out on a double-cantilever beam both at atmospheric pressure and in vacuum at a pressure of 1 mm of Hg and a temperature of 25°C. Figures 3.24 and 3.25 show these results.

The effect of the vibration frequency on the damping behavior of CFRP and GFRP was studied in Reference [50]. The authors compared the results obtained with the data for the other materials. Figure 3.26 displays a diagram from Reference [50].

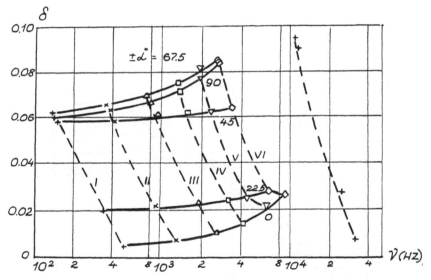

FIGURE 3.12. Logarithmic decrement of vibrations, δ, versus a frequency for six transverse modes and one longitudinal vibration mode for BFRP of five reinforced structures. Nomenclature: +, ×, △, □, ▽, and ◇ correspond to the first, second, third, fourth, fifth, and sixth specimen vibration modes respectively; the characteristics in longitudinal vibrations are marked by the symbol, +(separate curve).

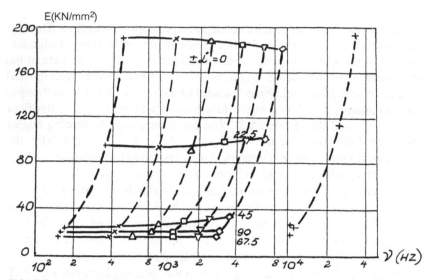

FIGURE 3.13. Effective Young's modulus, E, versus a frequency for six transverse vibration modes and one longitudinal vibration mode for BFRP of five reinforced structures. Nomenclature: see Figure 3.12.

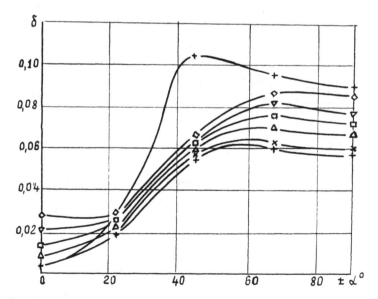

FIGURE 3.14. Logarithmic decrement of vibrations, δ, versus fiber orientation angle with respect to the specimen axis for six transverse vibration modes and one longitudinal vibration mode for BFRP of five reinforced structures. Nomenclature: see Figure 3.12.

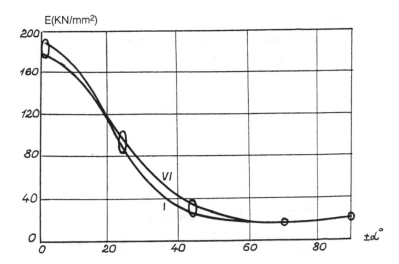

FIGURE 3.15. Effective Young's modulus, E, versus fiber orientation angle with respect to the specimen axis for six transverse vibration modes for BFRP of five reinforced structures.

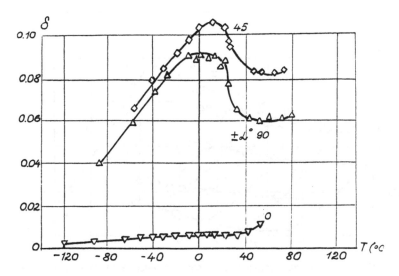

FIGURE 3.16. Variation of logarithmic decrement of vibrations, δ, with temperature for longitudinal vibration mode for BFRP of three reinforced structures: 0°, ±45°, and 90°.

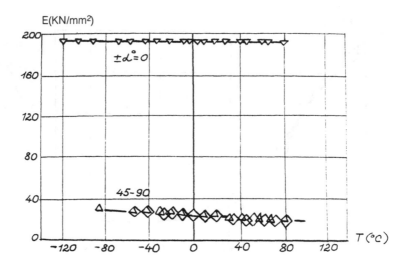

FIGURE 3.17. Variation of effective Young's modulus, E, with temperature for BFRP of three structures.

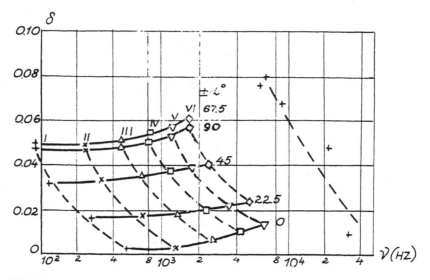

FIGURE 3.18. Logarithmic decrement of vibrations, δ, versus a frequency for six transverse vibration modes and one longitudinal vibration mode for CFRP of five reinforced structures. Nomenclature: see Figure 3.12.

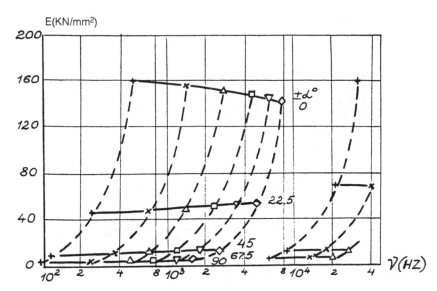

FIGURE 3.19. Young's modulus, E, versus a frequency for six transverse vibration modes and one longitudinal vibration mode for CFRP of five reinforced structures. Nomenclature: see Figure 3.12.

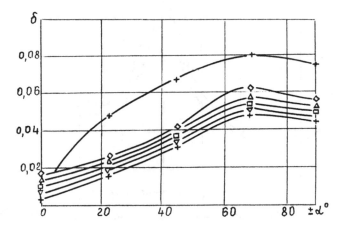

FIGURE 3.20. Logarithmic decrement of vibrations, δ, versus fiber orientation angle with respect to the specimen axis for five transverse vibration modes and one longitudinal vibration mode for CFRP. Nomenclature: see Figure 3.12.

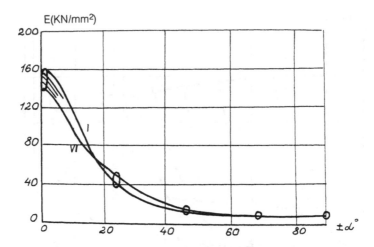

FIGURE 3.21. Young's modulus, E, versus fiber orientation angle with respect to the specimen axis for six transverse vibration modes for CFRP.

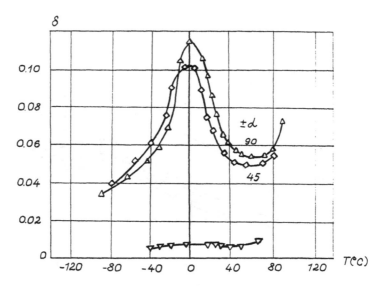

FIGURE 3.22. Logarithmic decrement of vibrations, δ, with temperature for longitudinal vibration mode for CFRP of three structures: 0°, ±45°, and 90°.

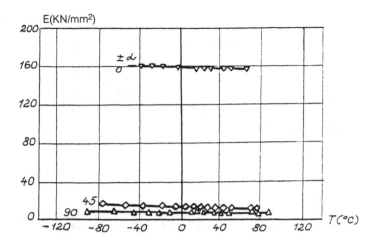

FIGURE 3.23. Variation of Young's modulus, E, with temperature for longitudinal vibration mode for CFRP of three structures.

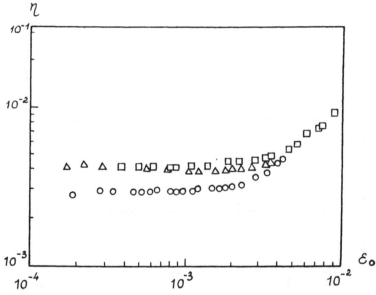

FIGURE 3.24. Relationship between loss factor, η, and amplitude of maximum strain, ϵ_0, of cantilever specimens for the first vibration mode in vacuum ($\bigcirc$), for the first ($\triangle$) and the second ($\square$) modes at atmospheric pressure. Material: cross-plied GFRP.

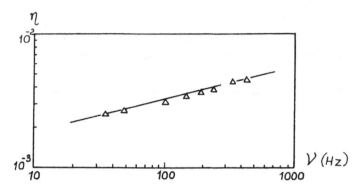

FIGURE 3.25. Relationship between loss factor, η, and vibration frequency, ν, of cantilever specimens for the first vibration mode in vacuum. Material: cross-plied GFRP.

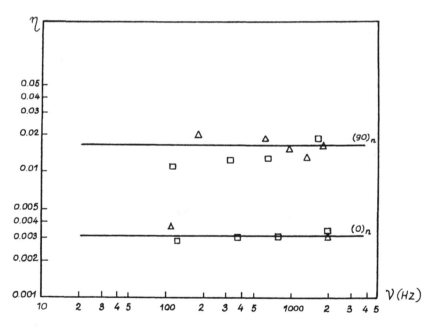

FIGURE 3.26. Variation of loss factor, η, with load frequency, ν, for unidirectional CFRP ($\triangle$) and GFRP ($\square$) with two fiber orientation angles, 0° and 90°.

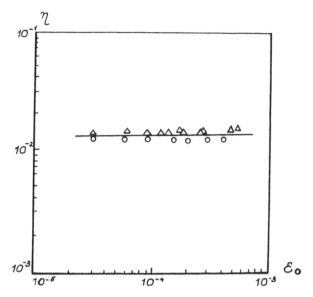

FIGURE 3.27. Loss factor, η, versus maximum strain amplitude, ϵ_0, of GFRP specimen with random orientation of chopped fibers; fiber volume fraction equals 65%. The first ($\triangle$) and the second ($\bigcirc$) vibration modes.

Reference [28] deals with the experimental investigation of the complex moduli of the composites with chopped fibers and of hybrid composites with oriented continuous fibers and a fraction of chopped fibers with random orientation. The authors examined the effect of the frequency and the load amplitude on the complex moduli of the specimens in the form of double-cantilever beams. Figure 3.27 illustrates the relationship between the loss factor and the maximum strain amplitude of the specimen. Double-cantilever specimens were tested in vacuum.

In conclusion, it should be said that the data discussed in the present chapter do not cover all the works in the field of experimental investigation of fibrous composite dissipative behavior and the references presented here are not a bibliographical review of the problem. Many of the works not mentioned in the chapter are included in the reference list. Here the authors tried to show only a part of the investigations reflecting the main directions in the experimental study of fibrous composite dissipative behavior.

CHAPTER 4

Elasto-Dissipative Characteristics (EDC) of Anisotropic Bodies

It is customary to use the term ''elastic'' to describe those solids for which stresses (σ_{ij}) are unambiguous functions of strains (ϵ_{ij}) during deformation, i.e.,

$$\sigma_{ij} = f(\epsilon_{ij}) \qquad (4.1)$$

Here σ_{ij} and ϵ_{ij} are the symmetrical stress and strain tensors of the second rank. Each of them is characterized by six independent components.

The behavior of elastic materials under loading and unloading is the same. The closed deformation path corresponds to the closed loading path. Work necessary for the closed deformation cycle is equal to zero.

The experiments show that many materials are linear elastic under small deformations, i.e., strains are proportional to applied stresses and vice versa. In the case of simple (uniaxial) deformation, this relation (Hooke's law) is written as

$$\sigma = E\epsilon$$

and

$$\epsilon = \sigma/E \qquad (4.2)$$

where E is the modulus of elasticity, the material constant.

Basically, elasticity is only the phenomenological model of the material behavior. It does not conform well to the real state of affairs. Real bodies, even under relatively small strains, exhibit a departure from ideal elasticity. The closed loading path appears to be infeasible without some irreversible energy expenditures that are damped by a body.

47

Energy losses in a unit volume of isotropic bodies in a complete loading cycle, ΔW, are usually believed to be proportional to the maximum value of the elastic energy, W, in a symmetric loading cycle

$$\Delta W = \psi W \qquad (4.3)$$

where ψ is the relative damping or the energy dissipation factor.

The energy method for consideration of the internal friction presupposes the introduction of an energy dissipation function in a complete loading cycle, provided that the stress-strain response remains linear elastic.

Anisotropic bodies, as objects with properties dependent upon coordinate orientation, have a more complex system of parameters that describe their elastic and dissipation behavior.

The present chapter considers: the hypothesis about the energy dissipation function for an anisotropic body, which depends on the amplitude of vibrations; special cases of material symmetry in anisotropic materials; the system of elastic and dissipation engineering constants and the rules of their transformation under coordinate system rotation.

It will be shown below that dissipative response of a body is closely related to its elastic response. Elastic properties determine the potential level of the damped energy. That is why it is reasonable to deal with the unified system of the elasto-dissipative characteristics, EDC.

The last section of the chapter discusses character weaknesses and errors which were typical for many early works concerning energy damping in anisotropic bodies and composites.

4.1 ELASTIC CHARACTERIZATION OF ANISOTROPIC BODIES

Experiments with anisotropic bodies show that when any stress tensor component (σ_{ij}) affects a body, appearance of any strain tensor (ϵ_{ij}) component is possible. Natural generalization of Hooke's law [Equation (4.2)] considering this condition, involves the assumption that strain tensor components are linear uniform functions of all stress tensor components (and vice versa)

$$\epsilon_{ij} = S_{ijkl}\sigma_{kl} \qquad (4.4)$$

$$\sigma_{ij} = G_{ijkl}\epsilon_{kl} \qquad (4.5)$$

$$i,j,k,l = 1,2,3$$

Here S_{ijkl} and G_{ijkl} are the compliance and stiffness coefficients of the material.

In accordance with the indirect test for tensor character (see Appendix A) the compliance coefficients and the stiffnesses are the components of corresponding tensors of the fourth rank, i.e., the compliance tensor and the stiffness tensor of the material. In the general case the number of components of these tensors is equal to $3^4 = 81$.

The relationships of Equation (4.4) look like the following in the developed form

$$\epsilon_{11} = S_{1111}\sigma_{11} + S_{1112}\sigma_{12} + S_{1113}\sigma_{13} + S_{1121}\sigma_{21}$$

$$+ S_{1122}\sigma_{22} + S_{1123}\sigma_{23} + S_{1131}\sigma_{31} + S_{1132}\sigma_{32} + S_{1133}\sigma_{33}$$

$$\epsilon_{12} = S_{1211}\sigma_{11} + S_{1212}\sigma_{12} + S_{1213}\sigma_{13} + S_{1221}\sigma_{21}$$

$$+ S_{1222}\sigma_{22} + S_{1223}\sigma_{23} + S_{1231}\sigma_{31} + S_{1232}\sigma_{32} + S_{1233}\sigma_{33} \qquad (4.6)$$

$$\cdots\cdots\cdots\cdots\cdots\cdots\cdots\cdots\cdots$$

$$\epsilon_{33} = S_{3311}\sigma_{11} + S_{3312}\sigma_{12} + S_{3313}\sigma_{13} + S_{3321}\sigma_{21}$$

$$+ S_{3322}\sigma_{22} + S_{3323}\sigma_{23} + S_{3331}\sigma_{31} + S_{3332}\sigma_{32} + S_{3333}\sigma_{33}$$

However, due to the symmetry of the tensors σ_{ij} and ϵ_{ij} (. . .) only six relations among nine in Equations (4.5) and (4.6) are independent. As a result, the number of independent components of the tensors G_{ijkl} and S_{ijkl} is reduced to thirty-six. It can be shown that the symmetry of these tensors with respect to the subscripts ij and kl takes place. This condition reduces the number of independent components to twenty-one.

Under coordinate system rotation, the coordinates of the fourth rank tensors, G_{ijkl} and S_{ijkl}, are transformed following the transformation rules (see Appendix A)

$$S'_{ijkl} = S_{pqrs}l_{ip}l_{jq}l_{kr}l_{ls}$$

$$G'_{ijkl} = G_{ijkl}l_{ip}l_{jq}l_{kr}l_{ls} \qquad (4.7)$$

where l_{ip} are the direction cosines of the angles between the axes of the new coordinate system, x_i', and the main coordinate system, x_p.

Potential energy of the elastic body can be written in one of the following equal forms

$$W = \frac{1}{2} \sigma_{ij}\epsilon_{ij}$$

$$W = \frac{1}{2} G_{ijkl}\epsilon_{ij}\epsilon_{kl} \tag{4.8}$$

$$W = \frac{1}{2} S_{ijkl}\sigma_{ij}\epsilon_{kl}$$

4.1.1 Matrix Form of Hooke's Law

The matrix form for stress and strain representation is commonly used in technical publications

$$\{\sigma\} = \begin{Bmatrix} \sigma_{11} \\ \sigma_{22} \\ \sigma_{33} \\ \sigma_{23} \\ \sigma_{13} \\ \sigma_{12} \end{Bmatrix} = \begin{Bmatrix} \sigma_1 \\ \sigma_2 \\ \sigma_3 \\ \tau_{23} \\ \tau_{13} \\ \tau_{12} \end{Bmatrix} \qquad \{\epsilon\} = \begin{Bmatrix} \epsilon_{11} \\ \epsilon_{22} \\ \epsilon_{33} \\ 2\epsilon_{23} \\ 2\epsilon_{13} \\ 2\epsilon_{12} \end{Bmatrix} = \begin{Bmatrix} \epsilon_1 \\ \epsilon_2 \\ \epsilon_3 \\ \gamma_{23} \\ \gamma_{13} \\ \gamma_{12} \end{Bmatrix} \tag{4.9}$$

Using matrix designations for stresses and strains, Equations (4.4) and (4.5) will be written in the matrix form as follows:

$$\{\epsilon\} = [S]\{\sigma\}$$

or

$$\begin{Bmatrix} \epsilon_1 \\ \epsilon_2 \\ \epsilon_3 \\ \gamma_{23} \\ \gamma_{13} \\ \gamma_{12} \end{Bmatrix} = \begin{bmatrix} S_{11} & S_{12} & S_{13} & S_{14} & S_{15} & S_{16} \\ S_{21} & S_{22} & S_{23} & S_{24} & S_{25} & S_{26} \\ S_{31} & S_{32} & S_{33} & S_{34} & S_{35} & S_{36} \\ S_{41} & S_{42} & S_{43} & S_{44} & S_{45} & S_{46} \\ S_{51} & S_{52} & S_{53} & S_{54} & S_{55} & S_{56} \\ S_{61} & S_{62} & S_{63} & S_{64} & S_{65} & S_{66} \end{bmatrix} \begin{Bmatrix} \sigma_1 \\ \sigma_2 \\ \sigma_3 \\ \tau_{23} \\ \tau_{13} \\ \tau_{12} \end{Bmatrix} \tag{4.10}$$

and

$$\{\sigma\} = [G]\{\epsilon\}$$

or

$$
\begin{Bmatrix} \sigma_1 \\ \sigma_2 \\ \sigma_3 \\ \tau_{23} \\ \tau_{13} \\ \tau_{12} \end{Bmatrix} = \begin{bmatrix} g_{11} & g_{12} & g_{13} & g_{14} & g_{15} & g_{16} \\ g_{21} & g_{22} & g_{23} & g_{24} & g_{25} & g_{26} \\ g_{31} & g_{32} & g_{33} & g_{34} & g_{35} & g_{36} \\ g_{41} & g_{42} & g_{43} & g_{44} & g_{45} & g_{46} \\ g_{51} & g_{52} & g_{53} & g_{54} & g_{55} & g_{56} \\ g_{61} & g_{62} & g_{63} & g_{64} & g_{65} & g_{66} \end{bmatrix} \begin{Bmatrix} \epsilon_1 \\ \epsilon_2 \\ \epsilon_3 \\ \gamma_{23} \\ \gamma_{13} \\ \gamma_{12} \end{Bmatrix}
\qquad (4.11)
$$

Here $[S]$ and $[G]$ are the symmetrical compliance and stiffness matrices with the dimension 6×6.

Changing from Equation (4.10) to Equation (4.11) is possible if the determinant

$$\det |S| \neq 0 \qquad (4.12)$$

The components of the matrices $[S]$ and $[G]$ with two subscripts are related to the components of the tensors S_{ijkl} and G_{ijkl} with four subscripts as follows:

$$S_{mn} = S_{ijkl}, \quad \text{if } m,n = 1,2,3;$$

$$S_{mn} = 2S_{ijkl}, \quad \text{if } m \text{ or } n \text{ is equal to } 4,5,6;$$

$$\qquad (4.13)$$

$$S_{mn} = 4S_{ijkl}, \quad \text{if } m \text{ and } n \text{ are equal to } 4,5,6;$$

$$G_{mn} = G_{ijkl}$$

Pair subscripts 11, 22, 33, 23 or 32, 31 or 13, 12 or 21 are changed for the subscripts 1, 2, 3, 4, 5, 6 in accordance with the following rule:

- ij is changed for the subscript i, if $i = j$.
- ij is changed for the subscript that is equal to $9 - i - j$, if $i \neq j$.

Potential energy of the elastic anisotropic body [Equation (4.8)] can be written in the matrix form as:

$$W = \frac{1}{2} \{\sigma\}^T \{\epsilon\} = \frac{1}{2} \{\epsilon\}^T \{\sigma\}$$

$$W = \frac{1}{2} \{\sigma\}^T [S] \{\sigma\} \qquad (4.14)$$

$$W = \frac{1}{2} \{\epsilon\}^T [G] \{\epsilon\}$$

Potential energy is a nonnegative value. Hence, if the components of stresses and strains are not identically equal to zero, the quadratic forms of Equation (4.14) are determined positively. It is necessary and sufficient that each of the following determinants of the matrices $[G]$ and $[S]$

$$g_{11}, \quad \begin{vmatrix} g_{11} & g_{12} \\ g_{21} & g_{22} \end{vmatrix}, \ldots, \det |G|$$

$$\tag{4.15}$$

$$s_{11}, \quad \begin{vmatrix} s_{11} & s_{12} \\ s_{21} & s_{22} \end{vmatrix}, \ldots, \det |S|$$

is positive. All principal minors of the determinants $\det |G|$ and $\det |S|$ are positive. This is why diagonal coefficients of the matrices $[G]$ and $[S]$ are also positive.

It follows from Equation (4.15), that

$$g_{11}g_{12} - g_{12}^2 > 0 \quad \text{and} \quad s_{11}s_{12} - s_{12}^2 > 0 \tag{4.16}$$

In the general case an inequality occurs, where

$$|g_{ij}| < |\sqrt{g_{ii}g_{jj}}| \text{ and } |s_{ij}| < |\sqrt{s_{ii}s_{jj}}| \tag{4.17}$$

4.1.2 Special Cases of Elastic Symmetry

Anisotropy of the general type, where the matrices of the compliance and stiffness coefficients in Equations (4.10) and (4.11) include twenty-one independent coefficients, is the rare case for real materials. Usually, the internal structure of the material is such that its elastic properties are the same in some directions. Then, the number of the independent coefficients in the matrices of compliance and stiffness coefficients can be greatly reduced. The form of Hooke's law becomes simpler, when selecting the proper coordinate system.

Let us consider some main cases of elastic symmetry.

4.1.2.1 THE PLANE OF ELASTIC SYMMETRY

If the elastic response of the anisotropic body is identical in two directions that are symmetrical with respect to some plane, then this plane is called the plane of elastic symmetry. In this case there are thirteen independent coefficients that describe the material's behavior [45]. Hooke's law is of the simpler form when one of the coordinate planes coincides with the elastic symmetry plane. For instance, if the (x_1,x_2) plane is the symmetry plane, the matrix of compliance coefficients in Equation (4.10) can be represented in the following form:

$$[S] = \begin{bmatrix} s_{11} & s_{12} & s_{13} & 0 & 0 & s_{16} \\ & s_{22} & s_{23} & 0 & 0 & s_{26} \\ & & s_{33} & 0 & 0 & s_{36} \\ & & & s_{44} & s_{45} & 0 \\ \text{sim.} & & & & s_{55} & 0 \\ & & & & & s_{66} \end{bmatrix} \qquad (4.18)$$

Specifically, in uniaxial tension of the anisotropic material in the direction perpendicular to the elastic symmetry plane, one obtains

$$\epsilon_1 = s_{13}\sigma_3 \quad \epsilon_2 = s_{23}\sigma_3 \quad \epsilon_3 = s_{33}\sigma_3$$

$$\gamma_{23} = 0 \quad \gamma_{13} = 0 \quad \gamma_{12} = s_{36}\sigma_3$$

The material is subjected only to shear in the (x_1,x_2) plane under such a stress state. It follows from Equation (4.18) that if one of the main axes of the stress state is perpendicular to the elastic symmetry plane, then one of the main axes of the strain state is also perpendicular to this plane. That is why the direction perpendicular to the elastic symmetry plane is called the main direction (or the main axis) of elasticity.

4.1.2.2 ORTHOTROPIC BODY

A body with three mutually perpendicular planes of elastic symmetry is called an orthotropic body. An orthotropic body has nine independent coefficients that describe its elastic behavior [45].

When coordinate planes are the planes of elastic symmetry, Hooke's law for the orthotropic body will take the simplest form; then the compliance matrix takes the form:

$$[S] = \begin{bmatrix} s_{11} & s_{12} & s_{13} & 0 & 0 & 0 \\ & s_{22} & s_{23} & 0 & 0 & 0 \\ & & s_{33} & 0 & 0 & 0 \\ & & & s_{44} & 0 & 0 \\ \text{sim.} & & & & s_{55} & 0 \\ & & & & & s_{66} \end{bmatrix} \qquad (4.19)$$

Compliance and stiffness coefficients become more obvious and are more convenient to determine from simple experiments if they are expressed in terms of the so-called engineering constants—moduli of elasticity, shear moduli, and Poisson's ratios. Then, instead of Equation (4.13), one obtains

$$[S] = \begin{bmatrix} 1/E_1 & -\nu_{21}/E_2 & -\nu_{31}/E_3 & 0 & 0 & 0 \\ -\nu_{12}/E_1 & 1/E_2 & -\nu_{32}/E_3 & 0 & 0 & 0 \\ -\nu_{13}/E_1 & -\nu_{23}/E_2 & 1/E_3 & 0 & 0 & 0 \\ 0 & 0 & 0 & 1/G_{23} & 0 & 0 \\ 0 & 0 & 0 & 0 & 1/G_{13} & 0 \\ 0 & 0 & 0 & 0 & 0 & 1/G_{12} \end{bmatrix} \quad (4.20)$$

Here E_1, E_2, and E_3 are the moduli of elasticity in the corresponding directions; G_{12}, G_{13}, and G_{23} are the shear moduli in the (x_1, x_2), (x_1, x_3), and (x_2, x_3) planes respectively; ν_{ij} are the Poisson's ratios, where the first subscript indicates the direction of the stress and the second subscript indicates the direction of the transverse deformation.

Certainly, the number of independent coefficients in Equation (4.20) remains equal to nine, as the following equalities are true due to the symmetry of the matrix of compliance coefficients:

$$E_2\nu_{21} = E_2\nu_{12}, \quad E_2\nu_{32} = E_3\nu_{23}, \quad E_3\nu_{13} = E_1\nu_{31} \quad (4.21)$$

In turn, the following relations follow from Equation (4.21)

$$\nu_{12}\nu_{23}\nu_{31} = \nu_{21}\nu_{32}\nu_{13}$$

Hooke's law in the form solved for stresses results in Equation (4.11). Applying engineering constants, one will obtain

$$g_{12} = E_1(1 - \nu_{23}\nu_{32})/\Delta$$

$$g_{22} = E_2(1 - \nu_{31}\nu_{13})/\Delta$$

$$g_{33} = E_3(1 - \nu_{12}\nu_{21})/\Delta$$

$$g_{12} = E_2(\nu_{12} + \nu_{13}\nu_{32})/\Delta = E_1(\nu_{21} + \nu_{23}\nu_{31})/\Delta \quad (4.22)$$

$$g_{13} = E_3(\nu_{13} + \nu_{12}\nu_{23})/\Delta = E_1(\nu_{31} + \nu_{32}\nu_{21})/\Delta$$

$$g_{23} = E_3(\nu_{23} + \nu_{21}\nu_{13})/\Delta = E_2(\nu_{32} + \nu_{31}\nu_{12})/\Delta$$

$$g_{44} = G_{23}, \quad g_{55} = G_{13}, \quad g_{66} = G_{12}$$

where

$$\Delta = \begin{vmatrix} 1 & -\nu_{12} & -\nu_{13} \\ -\nu_{21} & 1 & -\nu_{23} \\ -\nu_{31} & -\nu_{32} & 1 \end{vmatrix} = 1 - \nu_{12}\nu_{21} - \nu_{23}\nu_{32} - \nu_{31}\nu_{13} - 2\nu_{12}\nu_{23}\nu_{31}$$

4.1.2.3 THE PLANE OF ISOTROPY AND THE TRANSVERSELY ISOTROPIC (MONOTROPIC) BODY

A plane, where all directions are equivalent with respect to elastic properties, is called the plane of isotropy (isotropic plane). A body that has such a plane of isotropy is called the transversely isotropic body. Once the coordinate plane (x_2, x_3) coincides with the plane of isotropy, one can derive the following equation for this specific case of the orthotropic material

$$E_2 = E_3, \quad G_{13} = G_{12}$$

$$\nu_{12} = \nu_{13}, \quad \nu_{21} = \nu_{31}, \quad \nu_{23} = \nu_{32} \qquad (4.23)$$

$$G_{23} = E_2/2(1 + \nu_{23}) = E_3/2(1 + \nu_{32})$$

The compliance matrix is of the form

$$[S] = \begin{bmatrix} s_{11} & s_{12} & s_{12} & 0 & 0 & 0 \\ & s_{22}^0 & s_{23}^0 & 0 & 0 & 0 \\ & & s_{22}^0 & 0 & 0 & 0 \\ & & & s_{44} & 0 & 0 \\ \text{sim.} & & & & s_{55} & 0 \\ & & & & & s_{55} \end{bmatrix} \qquad (4.24)$$

or, in terms of engineering constants

$$[S] = \begin{bmatrix} 1/E_1 & -\nu_{21}/E_2 & -\nu_{21}/E_2 & 0 & 0 & 0 \\ -\nu_{12}/E_1 & 1/E_2 & -\nu_{32}/E_2 & 0 & 0 & 0 \\ -\nu_{12}/E_1 & -\nu_{23}/E_2 & 1/E_2 & 0 & 0 & 0 \\ 0 & 0 & 0 & 1/G_{23} & 0 & 0 \\ 0 & 0 & 0 & 0 & 1/G_{12} & 0 \\ 0 & 0 & 0 & 0 & 0 & 1/G_{12} \end{bmatrix} \qquad (4.25)$$

Shear modulus in the plane of isotropy, G_{23}, according to Equation (4.23) is not independent and the general number of independent constants for such a body is equal to five.

4.1.3 Plane Stress State

It is customary to assume that the plane stress state is a specific case of the material stress state, when

$$\sigma_3 = 0, \quad \tau_{23} = 0, \quad \tau_{13} = 0 \tag{4.26}$$

In the general case of anisotropy, as is seen from Equations (4.10) and (4.11), the corresponding strain components are not equal to zero

$$\epsilon_3 \neq 0, \quad \gamma_{23} \neq 0, \quad \gamma_{13} \neq 0 \tag{4.27}$$

Of particular interest is the case of the plane stress state for some specific types of material symmetry.

Thus, it is seen from Equations (4.14), (4.20), (4.22), (4.24), and (4.25), that in the case of the plane stress state Hooke's relations for orthotropic and transversely isotropic materials are of the same form

$$\begin{Bmatrix} \sigma_1 \\ \sigma_2 \\ \tau_{12} \end{Bmatrix} = \begin{bmatrix} g_{11}^0 & g_{12}^0 & 0 \\ g_{12}^0 & g_{22}^0 & 0 \\ 0 & 0 & g_{66}^0 \end{bmatrix} \begin{Bmatrix} \epsilon_1 \\ \epsilon_2 \\ \gamma_{12} \end{Bmatrix} \tag{4.28}$$

or

$$\begin{Bmatrix} \epsilon_1 \\ \epsilon_2 \\ \gamma_{12} \end{Bmatrix} = \begin{bmatrix} s_{11}^0 & s_{12}^0 & 0 \\ s_{12}^0 & s_{22}^0 & 0 \\ 0 & 0 & s_{66}^0 \end{bmatrix} \begin{Bmatrix} \sigma_1 \\ \sigma_2 \\ \tau_{12} \end{Bmatrix} \tag{4.29}$$

When using engineering constants, the relationships between stiffness coefficients, g_{ij}^0, and compliance coefficients, s_{ij}^0, will be written as follows:

$$g_{11}^0 = \frac{E_1}{1 - \nu_{12}\nu_{21}} = \frac{s_{22}^0}{s_{11}^0 s_{22}^0 - (s_{12}^0)^2}$$

$$g_{22}^0 = \frac{E_2}{1 - \nu_{12}\nu_{21}} = \frac{s_{11}^0}{s_{11}^0 s_{22}^0 - (s_{12}^0)^2}$$

$$g_{12}^0 = \frac{\nu_{21}E_1}{1 - \nu_{12}\nu_{21}} = \frac{s_{12}^0}{s_{11}^0 s_{22}^0 - (s_{12}^0)^2}$$

$$g_{66}^0 = G_{12} = \frac{1}{s_{66}^0}$$

$$(4.30)$$

$$s_{11}^0 = \frac{1}{E_1} = \frac{g_{22}^0}{g_{11}^0 g_{22}^0 - (g_{12}^0)^2}$$

$$s_{22}^0 = \frac{1}{E_2} = \frac{g_{11}^0}{g_{11}^0 g_{22}^0 - (g_{12}^0)^2}$$

$$s_{12}^0 = \frac{\nu_{12}}{E_1} = \frac{g_{12}^0}{g_{11}^0 g_{22}^0 - (g_{12}^0)^2}$$

$$s_{66}^0 = \frac{1}{G_{12}} = \frac{1}{g_{66}^0}$$

Stiffness and compliance matrices have the structure of Equations (4.28) and (4.29) only in the case when the main axes of symmetry coincide with the coordinate system axes. In this case, matrix components are marked by the superscript zero.

At the arbitrary arrangement of coordinate axes the matrices [G] and [S] in Equations (4.28) and (4.29) can be completely full, i.e., they do not have terms that are equal to zero.

4.1.4 Transformation of Elastic Characteristic under Coordinate System Rotation

In the general case, transformation of stress and strain tensors with relation to new coordinate axes is realized in accordance with the well-known formulas (see Appendix A)

$$\sigma_{i'j'} = \sigma_{ij} l_{i'i} l_{j'j}$$

$$(4.31)$$

$$\epsilon_{i'j'} = \epsilon_{ij} l_{i'i} l_{j'j}$$

$$i', j' = 1,2,3, \quad i, j = 1,2,3$$

When the (x_1, x_2), axes rotate through the angle α, the direction cosines, l_{ip} (in the matrix form), are of the form

$$[l] = \begin{bmatrix} l_{11} & l_{12} & l_{13} \\ l_{21} & l_{22} & l_{23} \\ l_{31} & l_{32} & l_{33} \end{bmatrix} = \begin{bmatrix} \cos\alpha & -\sin\alpha & 0 \\ \sin\alpha & \cos\alpha & 0 \\ 0 & 0 & 1 \end{bmatrix}$$

$$(4.32)$$

Using Equations (4.31) and (4.32) for the case of the plane stress state, the relations for stress transformation are obtained in the form

$$\{\sigma'\} = [T_1]\{\sigma_{12}\} \tag{4.33}$$

or

$$\begin{Bmatrix} \sigma_1' \\ \sigma_2' \\ \tau_{12}' \end{Bmatrix} = \begin{bmatrix} c^2 & s^2 & -2\,sc \\ s^2 & c^2 & 2\,sc \\ sc & -sc & c^2 - s^2 \end{bmatrix} \begin{Bmatrix} \sigma_1 \\ \sigma_2 \\ \tau_{12} \end{Bmatrix}$$

Here $s = \sin\alpha$, $c = \cos\alpha$.

Equation (4.31) can be applied to the tensor components. Using the relationship between matrix and tensor designations for strains [Equation (4.9)] one can write

$$\{\epsilon\} = \begin{Bmatrix} \epsilon_1 \\ \epsilon_2 \\ \gamma_{12} \end{Bmatrix} = [\Omega] \begin{Bmatrix} \epsilon_{11} \\ \epsilon_{22} \\ \epsilon_{12} \end{Bmatrix} \tag{4.34}$$

where $[\Omega] = \lceil 1,1,2 \rfloor$ is the diagonal matrix.

Then, the tensor components of stresses relate to the matrix ones by an inverse relationship

$$\begin{Bmatrix} \epsilon_{11} \\ \epsilon_{22} \\ \epsilon_{12} \end{Bmatrix} = [\Omega]^{-1} \begin{Bmatrix} \epsilon_1 \\ \epsilon_2 \\ \gamma_{12} \end{Bmatrix} \tag{4.35}$$

Applying Equations (4.31) and (4.35), one can obtain

$$[\Omega]^{-1}\{\epsilon'\} = [T_1][\Omega]^{-1}\{\epsilon\} \tag{4.36}$$

Then,

$$\{\epsilon'\} = [\Omega][T_1][\Omega]^{-1}\{\epsilon\}$$

or

$$\{\epsilon'\} = [T_2]\{\epsilon\} \tag{4.37}$$

where $[T_2]$ is the transformation matrix for the matrix components of strains, which is of the form:

$$[T_2] = [\Omega][T_1][\Omega]^{-1} = \begin{bmatrix} c^2 & s^2 & -sc \\ s^2 & c^2 & sc \\ 2\,sc & -2\,sc & c^2 - s^2 \end{bmatrix} \qquad (4.38)$$

The formulas of inverse transformations follow from Equations (4.33) and (4.37)

$$\{\sigma\} = [T_1]^{-1}\{\sigma'\} \qquad (4.39)$$

$$\{\epsilon\} = [T_2]^{-1}\{\epsilon'\} \qquad (4.40)$$

The matrices for inverse transformations are expressed as

$$[T_1]^{-1} = \begin{bmatrix} c^2 & s^2 & 2\,sc \\ s^2 & c^2 & -2\,sc \\ -sc & sc & c^2 - s^2 \end{bmatrix} \qquad (4.41)$$

$$[T_2]^{-1} = \begin{bmatrix} c^2 & s^2 & sc \\ s^2 & c^2 & -sc \\ -2\,sc & 2\,sc & c^2 - s^2 \end{bmatrix} \qquad (4.42)$$

The following identities are convenient to use, when performing the calculations

$$[T_1]^{-1} = [T_2]^{\mathrm{T}}, \quad [T_2]^{-1} = [T_1]^{\mathrm{T}}$$

$$([T_1]^{-1})^{\mathrm{T}} = [T_2], \quad ([T_2]^{-1})^{\mathrm{T}} = [T_1] \qquad (4.43)$$

$$[T_1]^{-1} = [T_1(-\alpha)], \quad [T_2]^{-1} = [T_2(-\alpha)]$$

Substituting Equations (4.39) and (4.40) into Equation (4.38), one can obtain

$$\{\sigma'\} = [\overline{G}]\{\epsilon'\} \qquad (4.44)$$

where the stiffness matrix in the new coordinate axes (x_1', x_2') is calculated in the following way:

$$[\overline{G}] = \begin{bmatrix} \overline{g}_{11} & \overline{g}_{12} & \overline{g}_{16} \\ & \overline{g}_{22} & \overline{g}_{26} \\ \text{sim.} & & \overline{g}_{66} \end{bmatrix} = [T_1][G^0][T_1]^{\mathrm{T}} \qquad (4.45)$$

Matrix components are of the form:

$$\bar{g}_{11} = c^4 g_{11}^0 + s^4 g_{22}^0 + 2 (g_{12}^0 + 2g_{66}^0) s^2c^2$$

$$\bar{g}_{12} = (g_{11}^0 + g_{22}^0 - 4g_{66}^0) s^2c^2 + (s^4 + c^4) g_{12}^0$$

$$\bar{g}_{16} = [c^2 g_{11}^0 - s^2 g_{22}^0 + (g_{12}^0 + 2g_{66}^0) (s^2 - c^2)] \, sc$$

$$\bar{g}_{22} = s^4 g_{11}^0 + c^4 g_{22}^0 + 2(g_{12}^0 + 2g_{66}^0) s^2c^2 \qquad (4.46)$$

$$\bar{g}_{26} = [s^2 g_{11}^0 - c^2 g_{22}^0 - (g_{12}^0 + 2g_{66}^0) (s^2 - c^2)] \, sc$$

$$\bar{g}_{66} = (g_{11}^0 - 2g_{12}^0 + g_{22}^0) s^2c^2 + (s^2 - c^2) g_{66}^0$$

The relations that are inverse to Equation (4.44) are given by:

$$\{\epsilon'\} = [\bar{S}]\{\sigma'\} \qquad (4.47)$$

where the compliance matrix $[\bar{S}]$ is calculated as follows

$$[\bar{S}] = [T_2][S^0][T_2]^T \qquad (4.48)$$

Its structure is analogous to the structure of the matrix $[\bar{G}]$ [see Equation (4.45)] and its components are of the form

$$\bar{s}_{11} = c^4 s_{11}^0 + s^4 s_{22}^0 + 2(s_{12}^0 + 2s_{66}^0) s^2c^2$$

$$\bar{s}_{12} = (s_{11}^0 + s_{22}^0 - 4s_{66}^0) s^2c^2 + (s^4 + c^4)s_{12}^0$$

$$\bar{s}_{16} = [2 c^2 s_{11}^0 - 2 s^2 s_{22}^0 + (s_{12} + s_{66}^0)(s^2 - c^2)] \, sc$$

$$\bar{s}_{22} = s^4 s_{11}^0 + c^4 s_{22}^0 + (2s_{12}^0 + s_{66}^0) s^2c^2 \qquad (4.49)$$

$$\bar{s}_{26} = [2 s^2 s_{11}^0 - 2 c^2 s_{22}^0 - (2s_{12}^0 + s_{66}^0)(s^2 - c^2)] \, sc$$

$$\bar{s}_{66} = (4s_{11}^0 - 8s_{12}^0 + 4s_{22}^0) s^2c^2 + (s^2 - c^2)s_{66}^0$$

4.2 ENERGY LOSSES UNDER CYCLIC LOADING OF AN ANISOTROPIC BODY

The following hypotheses are taken into consideration when describing the energy dissipation of an arbitrary anisotropic body under cyclic loading:

- Energy losses, ΔW, in a unit volume of a body in a complete loading cycle are defined by the function of stress and/or strain tensors and may depend upon body temperature T and loading frequency ω

$$\Delta W = f(\sigma_{ij}, \epsilon_{ij}, T, \omega) \qquad (4.50)$$

where the function f meets the necessary requirements of stress and strain Taylor expansion.

- Loading is carried out according to symmetric monoharmonic cycles so that σ_{ij} and ϵ_{ij} in Equation (4.50) and in analogous expressions below are the amplitude values of the components of stress and strain tensors during the loading cycle.

- Stresses and strains are connected by Hooke's law, i.e.,

$$\sigma_{ij} = G_{ijkl}\epsilon_{kl} \qquad (4.51)$$

or

$$\epsilon_{ij} = S_{ijkl}\sigma_{kl} \qquad (4.52)$$

where G_{ijkl} is the tensor of elastic moduli (the stiffness tensor), and S_{ijkl} is the compliance tensor.

Let us represent Equation (4.50) as a Taylor series according to the powers of σ_{ij} or ϵ_{ij} components

$$\Delta W = \frac{1}{2}(\Psi_{ij}\sigma_{ij} + \Psi_{ijkl}\sigma_{ij}\sigma_{kl} + \dots) \qquad (4.53)$$

$$\Delta W = \frac{1}{2}(\Phi_{ij}\epsilon_{ij} + \Phi_{ijkl}\epsilon_{ij}\epsilon_{kl} + \dots) \qquad (4.54)$$

Here the coefficients $\Psi_{ij}\,\Psi_{ijkl}, \dots$ and $\Phi_{ij}\,\Phi_{ijkl}, \dots$ are the tensors of the corresponding rank, as contractions of these values with stress (strain) tensors are scalars (see the indirect test for tensor character in Appendix A). In general, the components of these tensors depend on T and ω.

Usually the energy dissipation in solids is characterized by the relative energy dissipation (the dissipation factor) ψ which is defined as a ratio of energy losses, ΔW, in a unit volume of a body to the amplitude value of the elastic energy, W, under given stress-strain state:

$$\psi = \frac{\Delta W}{W} \qquad (4.55)$$

where the amplitude of the elastic energy in a unit volume of a body in a complete loading cycle is expressed by Equation (4.8).

Equations (4.53) and (4.54) must be invariant for the change of the signs of amplitude values to the opposite ones. That is why only the terms with even powers relating to stresses (strains) should remain in the series, i.e., EDC tensors of the fourth, eighth, twelfth, etc. ranks.

4.2.1 Amplitude-Independent Internal Friction

Let us consider the case of the amplitude-independent internal friction of an anisotropic body. Under proportional changing of amplitudes of all stress (strain) components $\sigma_{ij}' = \lambda\sigma_{ij}$, $\epsilon_{ij}' = \lambda\epsilon_{ij}$ in such a body (λ is a parameter), the dissipation factor does not depend on λ. The specific elastic energy, W', is the uniform stress (strain) function of the second power

$$W' = \frac{1}{2}\lambda^2 G_{ijkl}\epsilon_{ij}\epsilon_{kl} = \frac{1}{2}\lambda^2 S_{ijkl}\sigma_{ij}\sigma_{kl} = \lambda^2 W$$

The energy loss function in the unit volume of a body in a complete loading cycle [Equation (4.50)] also must be the uniform function of the second power. Then only quadratic terms remain in Equations (4.53) and (4.54). Thus, the amplitude-independent internal friction in the anisotropic body is described by the relations for energy losses in a unit volume in a loading cycle as follows:

$$\Delta W = \frac{1}{2}\Psi_{ijkl}\sigma_{ij}\sigma_{kl}$$

$$\Delta W = \frac{1}{2}\Phi_{ijkl}\epsilon_{ij}\epsilon_{kl}$$

(4.56)

Using Equations (4.51) and (4.52), let us represent Equation (4.56) in the mixed form:

$$\Delta W = \frac{1}{2}X_{ijkl}\epsilon_{ij}\sigma_{kl}$$

$$\Delta W = \frac{1}{2}X_{ijkl}^*\sigma_{ij}\epsilon_{kl}$$

(4.57)

The components Ψ_{ijkl}, Φ_{ijkl}, X_{ijkl} and X_{ijkl}^* form tensors of the fourth rank, as they give a scalar under contraction with two tensors of the second rank (σ_{ij} and ϵ_{ij}). They are related in the following way:

$$\Psi_{ijkl} = X_{mnkl}S_{mnij} = X^*_{ijmn}S_{mnkl} = \Phi_{mnps}S_{pskl}S_{mnij}$$

$$\Phi_{ijkl} = X_{ijmn}G_{mnkl} = X^*_{mnkl}G_{mnij} = \Psi_{psmn}G_{psij}G_{mnkl}$$

$$X_{ijkl} = \Psi_{mnkl}G_{mnij} = \Phi_{ijmn}S_{mnkl} = X^*_{klij}$$
(4.58)

$$X^*_{ijkl} = \Psi_{ijmn}G_{mnkl} = \Phi_{mnkl}S_{mnij} = X_{klij}$$

Tensors for elastic characteristics of the anisotropic body are in Equation (4.58). That is why we shall term the tensor, Ψ_{ijkl}, as the elasto-dissipative characteristic (EDC) tensor related to stresses or the elasto-dissipative characteristic stress tensor (EDC stress tensor), the tensor, Φ_{ijkl}, as the EDC tensor related to strains (EDC strain tensor), and X_{ijkl} and X^*_{ijkl} as the mixed EDC tensors of the anisotropic body.

The relations in Equation (4.56) are the quadratic forms relating to stress and strain tensor components. As the tensors Ψ_{ijkl} and Φ_{ijkl} must determine energy losses unambiguously, then the quadratic forms are the symmetric ones. In fact, there cannot be any difference between the coefficients, for example, relating to $\sigma_{ij}\sigma_{kl}$ and $\sigma_{kl}\sigma_{ij}$. At least their values always can be made equal. That is why the tensors Ψ_{ijkl} and Ψ_{ijkl} are symmetric regarding the transposition of the first and the second pairs of the subscripts

$$\Phi_{ijkl} = \Phi_{klij}, \quad \Psi_{ijkl} = \Psi_{klij}$$
(4.59)

It follows from the symmetry of the tensors σ_{ij} and ϵ_{ij} that the components of the EDC stress and strain tensors do not change when the subscript (i and j, k, and l) transposition takes place

$$\Phi_{ijkl} = \Phi_{jikl} = \Phi_{ijlk}, \quad \Psi_{ijkl} = \Psi_{jikl} = \Psi_{ijlk}$$
(4.60)

Equations (4.59) and (4.60) enable one to reduce the number of different components of the EDC tensors. Only twenty-one components among eighty-one of the fourth rank tensor remain different in three-dimensional (Ψ_{ijkl}, Φ_{ijkl}) space.

The symmetry for mixed EDC tensors [Equation (4.57)] takes place when transpositions are made in the subscripts i and j, k and l

$$X_{ijkl} = X_{jikl} = X_{ijlk} = X_{jilk}$$
(4.61)

EDC tensors are the tensors of the fourth rank. That is why, when exchanging Cartesian coordinates (x_1, x_2, x_3) for Cartesian coordinates

(x_1', x_2', x_3'), their components are transformed according to the tensor rule (see Appendix A). For example, the EDC stress tensor, Ψ_{ijkl}', in new coordinates is defined as follows:

$$\Psi_{ijkl}' = \Psi_{pqrs} l_{ip} l_{jq} l_{kr} l_{ls} \tag{4.62}$$

where l_{ip} are the direction cosines of the angles between x_i'- and x_p-axes.

In the expansions of energy losses [Equations (4.53) and (4.54)] for the case of the amplitude-dependent internal friction of the anisotropic body one may take into account the tensor coefficients of the higher ranks (the 8th, 12th, etc.). It should be admitted that consideration of the amplitude dependence of an anisotropic body's internal friction requires the determination of a rather large number of independent EDC tensor components. The number of independent components in the tensors of high ranks is given in Table 4.1 [45].

There is no need to explain that determining such a large number of the constants is a rather difficult problem in practice.

In the case of small stress amplitudes, dissipation factors for most anisotropic bodies are experimentally shown not to depend upon the amplitude. That is why further principal attention will be paid to the amplitude-independent internal friction in anisotropic bodies.

4.2.2 Amplitude-Dependent Internal Friction of Isotropic Bodies

The derivation of a model for energy losses in an isotropic body under amplitude-dependent internal friction is less cumbersome than it is for an anisotropic body. This makes it possible to consider the relationship between energy dissipation factors and the stress level under an arbitrary volume stress state.

Dissipation constants are defined from the approximation of empirical

TABLE 4.1. Independent Components of the Tensors.

Symmetry Type	Rank of the Tensor, $\Psi_{ijkl} \ldots (i,j,k,l = 1,2,3)$		
	4	8	12
Anisotropy of the general type	21	126	462
Orthotropic body	9	42	138
Transversely isotropic body	5	16	39
Isotropic body	2	4	7

relations between energy dissipation factors and the stress level under simple types of stress states.

Let us derive the basic relations for energy losses in the isotropic body, regardless of Equation (4.53), but with the same assumptions.

For an isotropic body $\Delta W = f(\sigma_{ij})$ does not depend generally upon a coordinate system transformation. Therefore, energy losses will be written as a function of three independent invariants of the stress tensor. The following invariants can be used:

$$\Sigma_{\text{I}} = \sigma_{ii}$$

$$\Sigma_{\text{II}} = \sigma_{ij}\sigma_{ij} \qquad (4.63)$$

$$\Sigma_{\text{III}} = \sigma_{ij}\sigma_{jk}\sigma_{ki}$$

or

$$I_1 = \Sigma_{\text{I}}$$

$$I_2 = \frac{1}{2}(\Sigma_{\text{I}}^2 - \Sigma_{\text{II}}) \qquad (4.64)$$

$$I_3 = \frac{1}{6}(\Sigma_{\text{I}}^3 - 3\Sigma_{\text{I}}\Sigma_{\text{II}} + 2\Sigma_{\text{III}})$$

Now, let us present the expression for energy losses as a series relating to the powers of the invariants I_1, I_2, and I_3. It is known that $\psi = \Delta W/W$ cannot change when changing the stress signs to the opposite ones. Therefore, only the terms with stress powers that are multiples of two should be retained in the expression for ΔW

$$\Delta W = \frac{1}{2}(A_2 I_1^2 + B_2 I_2 + A_4 I_1^4 + B_4 I_2^2$$

$$+ A_6 I_1^6 + B_6 I_2^3 + C_6 I_3^2 + \ldots) \qquad (4.65)$$

Equation (4.65) defines energy losses of an isotropic body under an arbitrary volume stress state. Here the coefficients A_i, B_i, and C_i correspond to the invariants I_1, I_2, and I_3. They can depend on the temperature, T, and the loading frequency, ω. The subscripts of the constants A, B, and C conform to the power of stress for the term of the series expansion. If the expansion is constrained only by the sixth power of stresses, then one will obtain the amplitude dependence of relative energy dissipation of about the fourth power for simple types of stress states.

A set of dissipation constants describes the amplitude-dependent internal friction; they are to be determined experimentally. There are seven constants: A_2, B_2, A_4, B_4, A_6, B_6, and C_6.

The expression for the amplitude value of the elastic energy of a unit volume of a body in terms of the stress tensor invariants will be written as:

$$W = \frac{1}{2E} [I_1^2 - 2(1 + \nu)I_2] \qquad (4.66)$$

where E is the Young's modulus, and ν is the Poisson's ratio of the isotropic body.

Let us remember that the model of amplitude-dependent energy damping is under consideration. To determine the values of seven dissipation constants of the isotropic body, it is necessary to carry out the experiments and to obtain the relationships between the relative energy dissipation, ψ, and the stress level. For example, it is sufficient to define the relations for uniaxial loading, ideal shear, and three-dimensional (3-D) hydrostatic loading.

Assume that the material is under uniaxial cyclic loading along the axis 1 by the stress, σ. The stress tensor will then take the form $\sigma_{11} = \sigma$, $\sigma_{ij} = 0$ ($ij \neq 11$), stress invariants are: $I_1 = \sigma$, $I_2 = 0$, $I_3 = 0$. The expressions for energy losses and the elastic energy [Equations (4.65) and (4.66)] will be written as

$$\Delta W(\sigma) = \frac{1}{2}(A_2\sigma^2 + A_4\sigma^4 + A_6\sigma^6)$$

$$W(\sigma) = \frac{1}{2E}\sigma^2 \qquad (4.67)$$

The relationship of the fourth power will be obtained for the dissipation factor, ψ_σ:

$$\psi_\sigma(\sigma) = E(A_2 + A_4\sigma^2 + A_6\sigma^4) \qquad (4.68)$$

It is necessary to approximate experimental data with the help of the polynom [Equation (4.68)] and to calculate the values of three dissipation constants A_2, A_4, and A_6.

Dissipation constants, B_i, can be defined from the experiments in ideal shear, for example in the plane of the axes (1,2). In this case $\sigma_{12} = \tau$, $\sigma_{ij} =$

0 ($ij \neq 12$, $ji \neq 12$). The dissipation factor under ideal shear will be designated as $\psi_r(\tau)$. Stress tensor invariants will be of the form: $I_1 = 0$, $I_2 = -\tau^2$, $I_3 = 0$. The expressions analogous to Equation (4.42) will then take the form for ideal shear:

$$\Delta W(\tau) = \frac{1}{2}(-B_2\tau^2 + B_4\tau^4 - B_6\tau^6)$$

$$\text{(4.69)}$$

$$W(\sigma) = \frac{1}{2G}\tau^2$$

where G is the shear modulus of the material, relating to the other elastic constants as

$$G = \frac{E}{2(1 + \nu)}$$

Dissipation factor $\psi_r(\tau)$ is represented in the form of the following relation

$$\psi_r(\tau) = G(-B_2 + B_4\tau^2 - B_6\tau^4) \qquad \text{(4.70)}$$

The values of three dissipation constants, B_2, B_4, and B_6, are determined from the approximation of experimental data by the relation for $\psi_r(\tau)$.

The last constant, C_6, can be defined from the experiments in 3-D hydrostatic loading. Assume that the corresponding dissipation factor, ψ_p, depends upon hydrostatic pressure, p. The stress tensor for this loading type is of the form

$$\sigma_{11} = \sigma_{22} = \sigma_{33} = p, \quad \sigma_{ij} = 0 \ (i \neq j)$$

Stress invariants take the following values

$$I_1 = 3p, \quad I_2 = 3p^2, \quad I_3 = p^3$$

As a result, the expressions for energy losses and the elastic energy are of the forms

$$\Delta W = \frac{1}{2}[(9A_2 + 3B_2)p^2 + (81A_4 + 9B_4)p^4 + (729A_6 + 27B_6 + C_6)p^6]$$

$$\text{(4.71)}$$

$$W = \frac{1}{2K}p^2$$

where $K = E/3(1 - 2\nu)$ is the bulk elastic modulus of the material.

The dissipation factor $\psi_p(p)$ is defined by the following formula that includes an unknown constant, C_6

$$\psi_p(p) = K[9A_2 + 3B_2 + (81A_4 + 9B_4)p^2 + (729A_6 + 27B_6 + C_6)p^4]$$

(4.72)

Approximating the experimental relation of dissipation factors with the help of Equation (4.72), one can obtain the value of the constant, C_6.

The experiments in 3-D cyclic loading for determining the relative energy losses may be a rather difficult technical problem. That is why the number of constants can be reduced in Equation (4.65), if the material is assumed not to dissipate energy under these loading conditions. Then, energy losses, ΔW, will be a function of the deviator of the stress tensor. It means for the isotropic body that ΔW is the function of two invariants of the deviator. The stress deviator, $\bar{S}_{ij}$, does not change when loading in 3-D tension or compression. It is defined as:

$$\bar{S}_{ij} = \sigma_{ij} - p\delta_{ij}$$

(4.73)

where the hydrostatic stress component is of the form:

$$p = \frac{1}{3}\sigma_{ii} = \frac{1}{3}\Sigma_{I} = \frac{1}{3}I_1$$

δ_{ij} is the Kronecker delta.

The invariants of the stress tensor deviator, $\bar{\Sigma}_I, \bar{\Sigma}_{II}, \bar{\Sigma}_{III}$, are equal respectively

$$\bar{\Sigma}_I = 0$$

$$\bar{\Sigma}_{II} = (\sigma_{ij} - \delta_{ij}p)(\sigma_{ij} - \delta_{ij}p)$$

$$\bar{\Sigma}_{III} = (\sigma_{ij} - \delta_{ij}p)(\sigma_{ik} - \delta_{ik}p)(\sigma_{kj} - \delta_{kj}p)$$

or in terms of the main stresses, $\sigma_1, \sigma_2, \sigma_3$

$$\bar{\Sigma}_{II} = \frac{1}{3}[(\sigma_1 - \sigma_2)^2 + (\sigma_2 - \sigma_3)^2 + (\sigma_3 - \sigma_1)^2]$$

(4.74)

$$\bar{\Sigma}_{III} = (\sigma_1 - \sigma_3)^3 + (\sigma_2 - \sigma_1)^3 + (\sigma_3 - \sigma_2)^3$$

The expression for energy losses, ΔW, will be of the form

$$\Delta W = \frac{1}{2}\,(D_2\overline{\Sigma}_\mathrm{II} + D_4\overline{\Sigma}_\mathrm{II}^2 + D_6\overline{\Sigma}_\mathrm{II}^3 + F_6\overline{\Sigma}_\mathrm{III}^2 + \ldots) \qquad (4.75)$$

Here the dissipation constants, D_i and F_i, have subscripts that conform to the powers of stresses for the terms of the series expansion, which include these constants [analogous to Equation (4.65)].

If the series is limited by the terms of the sixth power related to stresses, then the dissipative response of the isotropic body will be described by four independent constants D_2, D_4, D_6, and F_6. The values of the constants are defined from two experiments in uniaxial loading and ideal shear.

Let us suppose that $\sigma_{11} = \sigma$, $\sigma_{ij} = 0$ ($ij \neq 11$), then the relations for the deviator invariants in terms of main stresses enable one to determine $\overline{\Sigma}_\mathrm{II} = 2/3\,(\sigma^2)$, $\overline{\Sigma}_\mathrm{III} = 0$. Once the necessary transformations are made, one will obtain the following expression for the dissipation factor under uniaxial cyclic loading:

$$\psi_\sigma(\sigma) = E\left(\frac{2}{3}D_2 + \frac{4}{9}D_4\sigma^2 + \frac{8}{27}D_4\sigma^4\right) \qquad (4.76)$$

The values of three dissipation constants, D_i, can be determined by approximation of experimental data with the polynom [Equation (4.76)].

The second experiment is ideal shear in the (x_1,x_2) plane. In this case $\sigma_{12} = \tau$, $\sigma_{ij} = 0$ ($ij \neq 12$, $ij \neq 21$). The main stresses in ideal shear will be: $\sigma_1 = -\tau$, $\sigma_2 = \tau$, $\sigma_3 = 0$. The equations for the invariants of the tensor deviator will be of the form: $\overline{\Sigma}_\mathrm{II} = 2\tau^2$, $\overline{\Sigma}_\mathrm{III} = 6\tau^3$.

The expressions for the dissipation factor, ψ_τ, under ideal shear will be as follows:

$$\psi_\tau(\tau) = G[2D_2 + 4D_4\tau^2 + (8D_6 + 36F_6)\,\tau^4] \qquad (4.77)$$

where only one dissipation constant, F_6, is unknown. Its value is determined by approximation of the experimental relationship between ψ_τ and τ with the polynom [Equation (4.77)].

Let us note that in some cases more complete correspondence of experimental data to the values of the constants D_2, D_4, D_6, and F_6 is required. Then any approximation method may be used.

To check the theoretical relations, Equations (4.65) and (4.75), experimentally, it is sufficient to compare experimental and theoretical values of the dissipation factors under complex types of stress states, provided that the necessary values of the dissipation constants of an isotropic body have already been determined under simple loading conditions.

4.3 COMPLEX MODULI OF ANISOTROPIC BODIES

A model that describes energy damping under cyclic loading is the model of the linear viscoelastic material. Viscoelastic properties of a material can be shown through the results of experiments in creep and stress relaxation. These experiments are performed to determine the law of either stresses or strains variation with time, provided that the values of stresses (or strains) are kept constant.

Many special relationships are proposed for the state equations that relate stresses and strains. A rather general form for the physical relationships of the linear theory of the viscoelastic body is the relationship of the following form:

$$\sigma_{ij}(t) = \int_{-\infty}^{t} G_{ijkl}(t - \tau) \, d\epsilon_{kl}(\tau) \qquad (4.78)$$

where $G_{ijkl}(t)$ is the tensor of the stress relaxation function, $\sigma_{ij}(t)$ and $\epsilon_{ij}(t)$ are the stress and strain tensors depending on time, t [19]. Equation (4.78) considers that the stresses at any fixed moment of time are determined by the whole history of strain changing up to this moment.

Viscoelastic relations are also represented in the form of the functional relationship of strains versus stress function:

$$\epsilon_{ij}(t) = \int_{-\infty}^{t} S_{ijkl}(t - \tau) \, d\sigma_{kl}(\tau) \qquad (4.79)$$

Here $S_{ijkl}(t)$ is the tensor of the creep function.

Laplace transformation makes it possible to derive the relations between the tensors – the relaxation functions and the creep functions [65].

Generally speaking, the viscoelastic responses of anisotropic bodies are completely described by a set of creep functions or a set of stress relaxation functions. Depending on the viscoelastic material symmetry, the number of independent creep or relaxation functions is reduced and is equal to the number of independent elastic constants. The form of creep or relaxation functions can be defined on the basis of experiments in creep or stress relaxation under simple types of loading.

Let us consider the cyclic loading of a body with the frequency, ω, and the amplitude of strains, ϵ_{kl}^{0}. The loading is carried out in accordance with the law:

$$\epsilon_{kl}^{*}(t) = \epsilon_{kl}^{0} e^{i\omega t}$$

where $\epsilon_{kl}^{*}(t)$ is the complex strain. Then, Equation (4.78) is transformed to the following form:

$$\sigma_{ij}^*(t) = i\omega\epsilon_{kl}^0 e^{i\omega t} \int_0^\infty G_{ijkl}(u)e^{-i\omega u}du \qquad (4.80)$$

or

$$\sigma_{ij}^* = G_{ijkl}^* \epsilon_{kl}^*$$

where $u = t - \tau$, σ_{ij}^* are the complex stresses, and G_{ijkl}^* are the complex moduli.

The complex moduli can be expanded into real and imaginary parts:

$$G_{ijkl}^* = G_{ijkl}' + i\, G_{ijkl}''$$

where

$$G_{ijkl}'(\omega) = \omega \int_0^\infty G_{ijkl}(u) \sin(\omega u)du$$

$$\qquad (4.81)$$

$$G_{ijkl}''(\omega) = \omega \int_0^\infty G_{ijkl}(u) \cos(\omega u)du$$

Complex stresses and strains are related by the complex moduli in Hooke's law. That is why, when solving dynamic problems on the harmonic loading of the viscoelastic body, it is sufficient to use corresponding solutions of the linear theory of elasticity. In doing so, the elastic moduli should be changed to the complex moduli of the viscoelastic body.

Equation (4.80) can be considered as the Fourier transform of the creep function (see Appendix C). This is the inverse Fourier transform which enables one to find the original by its image. That is why the tensor of creep functions, $G_{ijkl}(t)$, can be restored from the complex moduli functions G_{ijkl}' (ω) or G_{ijkl}'' (ω) [65].

$$G_{ijkl}(t) = \frac{2}{\pi} \int_0^\infty \frac{G_{ijkl}'(\omega)}{\omega} \sin \omega t\, d\omega \qquad (4.82)$$

or

$$G_{ijkl}(t) = \frac{2}{\pi} \int_0^\infty \frac{G_{ijkl}''(\omega)}{\omega} \cos \omega t\, d\omega$$

Note that Equation (4.80) can be represented in the following form:

$$\sigma_{ij}^*(t) = \sigma_{ij}^0 e^{i(\omega t + \varphi_{ij})} \quad \text{(no summing!)} \qquad (4.83)$$

where $\sigma_{ij}^0 = |\sigma_{ijkl}^* \epsilon_{kl}^0|$, loss tangent tg $\varphi_{ij}(\omega) = [G_{ijkl}''(\omega)\epsilon_{kl}^0]/[G_{ijkl}'(\omega)\epsilon_{kl}^0]$ (do not sum over $i,j!$), the angles φ_{ij} are not the tensor objects, the angles are lag angles of the strain $\epsilon_{ij}(t) = \epsilon_{ij}^0 e^{i\omega t}$ with respect to the stress under stable harmonic vibrations.

Energy losses in a unit volume of a body in a complete loading cycle are defined as follows:

$$\Delta W = \oint \sigma_{ij} d\epsilon_{ij} = \oint \text{Re}\sigma_{ij}^* \, d(\text{Re}\epsilon_{ij}^*) \qquad (4.84)$$

It follows from Equations (4.80) and (4.81) that

$$\text{Re}\epsilon_{ij}^* = \epsilon_{ij}^0 \cos \omega t$$

$$\text{Re}\sigma_{ij}^* = \epsilon_{kl}^0 (G_{ijkl}' \cos \omega t - G_{ijkl}'' \sin \omega t)$$

Substituting the values of the last expressions and performing the integration in a loading period, one can obtain

$$\Delta W = \int_0^{2\pi/\omega} \omega (G_{ijkl}'' \sin^2 \omega t - \frac{1}{2} G_{ijkl}' \sin 2\omega t)\epsilon_{kl}^0 \epsilon_{ij}^0 \, dt \qquad (4.85)$$

or

$$\Delta W = \pi G_{ijkl}'' \epsilon_{ij}^0 \epsilon_{kl}^0$$

Comparison of Equations (4.85) and (4.56) shows that EDC strain tensor and the imaginary part of the tensor of complex moduli coincide with an accuracy of the constant factor 2π.

Let us note that the model of the elastic body with dissipation behavior, as opposed to the viscoelastic body, makes it possible to maintain the simplicity for determining stress-strain state. The model enables one to consider the amplitude dependence of relative energy dissipation and to describe frequency-independent internal friction. This is essential for some materials.

4.4 EDC TENSORS FOR THE MATERIALS WITH DIFFERENT TYPES OF SYMMETRY

Let us write Equations (4.56) and (4.57) in the matrix form (see Appendix B):

$$\Delta W = \frac{1}{2} \{\sigma\}^T [\Psi]\{\sigma\}$$

$$\Delta W = \frac{1}{2} \{\epsilon\}^T [X]\{\sigma\}$$

$$\Delta W = \frac{1}{2} \{\sigma\}^T [X^*]\{\epsilon\} \tag{4.86}$$

$$\Delta W = \frac{1}{2} \{\epsilon\}^T [\Phi]\{\epsilon\}$$

Here a set of the components for each of EDC stress and strain tensors, and mixed tensors is represented by the matrices $[\Psi]$, $[\Phi]$, $[X]$, and $[X^*]$ respectively.

An anisotropic body may have some types of symmetry. In that case the selection of the proper coordinate system results in a simple expression for the matrices $[\Psi]$, $[\Phi]$, $[X]$, and $[X^*]$. For specific cases of the material symmetry it can be shown (in the same order as it was done for the tensors of elastic moduli or compliance tensors) that the matrices $[\Psi]$ and $[\Phi]$ have structures that are similar to the structures of the matrices corresponding to the tensors G_{ijkl} and S_{ijkl} and the same number of independent constants.

The orthotropic body has three mutually orthogonal symmetry planes. Its elasto-dissipative response is defined by nine components of the EDC matrix. The matrix $[\Psi]$ is of the form:

$$[\Psi] = \begin{bmatrix} \psi_{11} & \psi_{12} & \psi_{13} & 0 & 0 & 0 \\ & \psi_{22} & \psi_{23} & 0 & 0 & 0 \\ & & \psi_{33} & 0 & 0 & 0 \\ & & & \psi_{44} & 0 & 0 \\ \text{sim.} & & & & \psi_{55} & 0 \\ & & & & & \psi_{66} \end{bmatrix} \tag{4.87}$$

The EDC matrix of a transversely isotropic body [for which the (x_1,x_2) plane is the symmetry plane] contains five independent components

$$[\Psi] = \begin{bmatrix} \psi_{11} & \psi_{12} & \psi_{13} & 0 & 0 & 0 \\ & \psi_{22} & \psi_{23} & 0 & 0 & 0 \\ & & \psi_{22} & 0 & 0 & 0 \\ & & & 2(\psi_{22} - \psi_{23}) & 0 & 0 \\ \text{sim.} & & & & \psi_{55} & 0 \\ & & & & & \psi_{66} \end{bmatrix} \tag{4.88}$$

4.5 SYSTEM OF ENGINEERING CONSTANTS

EDC matrices determine energy losses in the anisotropic body under an arbitrary stress state. The values of their components can be obtained from a set of material tests under the cyclic loading. In practice it is more convenient to determine the components of the EDC stress matrix [Ψ].

Thus, for an orthotropic material the values of dissipation factors ψ_1^*, ψ_2^*, and ψ_3^* [Equation (4.55)] can be determined, if the uniaxial loading along the x_1, x_2, and x_3 coordinate axes is carried out. From shear experiments in the (x_2, x_3), (x_1, x_3), and (x_1, x_2) coordinate planes it is possible to evaluate dissipation factors ψ_4^*, ψ_5^*, and ψ_6^*. A set of six experiments permits one to define the values of diagonal components of the matrix [Ψ], [Equation (4.87)].

$$\psi_{11} = \frac{\psi_1^*}{E_1}, \; \psi_{22} = \frac{\psi_2^*}{E_2}, \; \psi_{33} = \frac{\psi_3^*}{E_3}, \; \psi_{44} = \frac{\psi_4^*}{G_{23}}, \; \psi_{55} = \frac{\psi_5^*}{G_{13}}, \; \psi_{66} = \frac{\psi_6^*}{G_{12}} \quad (4.89)$$

Here E_1, E_2, and E_3 are Young's moduli under uniaxial loading along x_1, x_2, and x_3 axes; while G_{23}, G_{13}, and G_{12} are shear moduli in corresponding coordinate planes.

Let us denote dissipation factors determined in the experiments on biaxial uniform cyclic loading along x_1- and x_2-, x_1- and x_3-, x_2- and x_3-axes by ψ_{12}^*, ψ_{13}^*, and ψ_{23}^*. Then the rest of the [Ψ] matrix components [Equation (4.87)] is expressed as follows:

$$\psi_{12} = \frac{1}{2} \left\{ \frac{1}{E_1} [\psi_{12}^*(1 - 2\nu_{12}) - \psi_1^*] + \frac{1}{E_2} (\psi_{12}^* - \psi_2^*) \right\}$$

$$\psi_{13} = \frac{1}{2} \left\{ \frac{1}{E_1} [\psi_{13}^*(1 - 2\nu_{13}) - \psi_1^*] + \frac{1}{E_3} (\psi_{13}^* - \psi_3^*) \right\} \quad (4.90)$$

$$\psi_{23} = \frac{1}{2} \left\{ \frac{1}{E_2} [\psi_{23}^*(1 - 2\nu_{23}) - \psi_2^*] + \frac{1}{E_3} (\psi_{23}^* - \psi_3^*) \right\}$$

where ν_{12}, ν_{13}, and ν_{23} are the Poisson's ratios.

The total combination of nine factors can be assumed as the system of engineering constants, determining the dissipative behavior of the orthotropic body under an arbitrary stress state. These factors are the following:

- dissipation factors under uniaxial loading along the x_1-, x_2-, and x_3-axes — ψ_1^*, ψ_2^*, and ψ_3^*

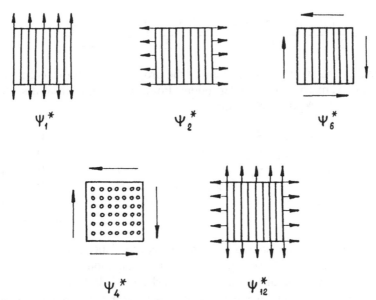

FIGURE 4.1. Schemes of experiments to determine the values of engineering dissipation constants of a transversely isotropic body.

- dissipation factors under shear in the (x_2,x_3), (x_1,x_3), and (x_1,x_2) coordinate planes $-\psi_4^*$, ψ_5^*, and ψ_6^*
- dissipation factors under uniform biaxial loading in coordinate planes along the coordinate axes $-\psi_{12}^*$, ψ_{13}^*, and ψ_{23}^*

Note that as related to the structure, $[\Psi]$ matrix components are the ratios of dissipation factors to elastic moduli, i.e., they depend upon both dissipation factors and material elastic constants. That is why it is advisable to term the $[\Psi]$, $[\Phi]$, $[X]$, and $[X^*]$ matrices as elasto-dissipative characteristic matrices of the anisotropic body.

The dissipative response of a transversely isotropic body is described by five coefficients [Equation (4.88)]. That is why it is enough to determine the following constants for such a body: ψ_1^*, ψ_2^*, ψ_4^*, ψ_6^*, and ψ_{12}^* (see Figure 4.1) [the (x_2,x_3) plane is the isotropic plane]. The values of $[\Psi]$ matrix components are calculated with the help of Equations (4.89) and (4.90), considering Equation (4.88). Realization of cyclic biaxial loading in order to estimate the coefficient, ψ_{12}^*, may involve some difficulties. That is why it is advantageous to change this procedure for uniaxial loading at the 45° angle to the x_1-axes in the (x_1,x_3) plane. The dissipation factor in this loading case will be designated as $\psi_{1/2}^*$. Then the value of the ψ_{12} component of the matrix $[\Psi]$ [Equation (4.88)] is calculated from the formula:

$$\psi_{12} = \frac{1}{2} \left[\frac{\psi_{1/2}^* (1 - 2\nu_{12}) - \psi_1^*}{E_1} + \frac{\psi_{1/2}^* - \psi_2^*}{E_2} + \frac{\psi_{1/2}^* - \psi_6^*}{G_{12}} \right] \tag{4.91}$$

In the case of complete isotropy of the properties, only two components of the EDC matrix remain independent. For example, the elasto-dissipative stress matrix $[\Psi]$ is of the form:

$$[\Psi] = \begin{bmatrix} \psi_{11} & \psi_{12} & \psi_{12} & 0 & 0 & 0 \\ & \psi_{11} & \psi_{12} & 0 & 0 & 0 \\ & & \psi_{11} & 0 & 0 & 0 \\ & & & 2(\psi_{11} - \psi_{12}) & 0 & 0 \\ \text{sim.} & & & & 2(\psi_{11} - \psi_{12}) & 0 \\ & & & & & 2(\psi_{11} - \psi_{12}) \end{bmatrix}$$

It is convenient to use two constants as independent engineering dissipation constants of the isotropic body – the dissipation factor in uniaxial loading, ψ_1^*, and the dissipation factor in shear, ψ_6^*. If Young's modulus is designated as E, and the shear modulus as G, then the components of the $[\Psi]$ matrix are expressed in terms of the elastic and dissipation constants in the following form:

$$[\Psi] = \begin{bmatrix} \psi_1^*/E & \dfrac{\psi_1^*}{E} - \dfrac{\psi_6^*}{2G} & \dfrac{\psi_1^*}{E} - \dfrac{\psi_6^*}{2G} & 0 & 0 & 0 \\ & \psi_1^*/E & \dfrac{\psi_1^*}{E} - \dfrac{\psi_6^*}{2G} & 0 & 0 & 0 \\ & & \psi_1^*/E & 0 & 0 & 0 \\ & & & \psi_6^*/G & 0 & 0 \\ \text{sim.} & & & & \psi_6^*/G & 0 \\ & & & & & \psi_6^*/G \end{bmatrix}$$

4.6 PLANE STRESS STATE

Let us discuss the plane stress state of the transversely isotropic body. Let the (x_2, x_3) plane be the plane of isotropy (the isotropic plane). The stresses $\sigma_3 = \sigma_{23} = \sigma_{13} = 0$. Therefore it is sufficient to consider reduced column matrices of stresses and strains

$$\{\sigma\} = \begin{Bmatrix} \sigma_1 \\ \sigma_2 \\ \sigma_{12} \end{Bmatrix} = \begin{Bmatrix} \sigma_1 \\ \sigma_2 \\ \tau_{12} \end{Bmatrix}, \quad \{\epsilon\} = \begin{Bmatrix} \epsilon_1 \\ \epsilon_2 \\ 2\epsilon_1 \end{Bmatrix} = \begin{Bmatrix} \epsilon_1 \\ \epsilon_2 \\ \gamma_{12} \end{Bmatrix}$$

and the matrices of elasto-dissipative characteristic

$$[\psi] = \begin{bmatrix} \psi_{11} & \psi_{12} & 0 \\ \psi_{12} & \psi_{22} & 0 \\ 0 & 0 & \psi_{66} \end{bmatrix}, \; [\varphi] = \begin{bmatrix} \varphi_{11} & \varphi_{12} & 0 \\ \varphi_{12} & \varphi_{22} & 0 \\ 0 & 0 & \varphi_{66} \end{bmatrix}, \; [\chi] = \begin{bmatrix} \chi_{11} & \chi_{12} & 0 \\ \chi_{21} & \chi_{22} & 0 \\ 0 & 0 & \chi_{66} \end{bmatrix} \; (4.92)$$

Here τ_{12} is the shear stress and γ_{12} is the shear strain.

The expressions for energy losses in terms of EDC matrices under plane stress state are of the forms:

$$\Delta W = \frac{1}{2} \{\sigma\}^T [\psi] \{\sigma\}$$

$$\Delta W = \frac{1}{2} \{\epsilon\}^T [\chi] \{\sigma\}$$

$$\Delta W = \frac{1}{2} \{\sigma\}^T [\chi^*] \{\epsilon\}$$

$$\Delta W = \frac{1}{2} \{\epsilon\}^T [\varphi] \{\epsilon\}$$

(4.93)

Corresponding components of the EDC stress matrices for volume stress state [Equation (4.88)] and for plane stress state [Equation (4.92)] are identical, but this is not true for the components of EDC strain matrices $[\Phi]$ and $[\varphi]$, and mixed matrices $[X]$ and $[\chi]$. The relation between the components of the matrices can easily be determined. If it is supposed that $\sigma_3 = 0$ in Hooke's law for volume stress state, then the following expression is derived for the strain, ϵ_3:

$$\epsilon_3 = -\frac{g_{13}}{g_{33}} \epsilon_1 - \frac{g_{23}}{g_{33}} \epsilon_2$$

Substituting the ϵ_3 value into Equation (4.86) and transforming the similar terms, one can derive the following relationships for the components of the matrices $[X]$ and $[\chi]$

$$\chi_{11} = X_{11} - X_{31} g_{13}/g_{33}$$

$$\chi_{12} = X_{12} - X_{32} g_{13}/g_{33}$$

$$\chi_{21} = X_{21} - X_{31} g_{23}/g_{33}$$

(4.94)

$$\chi_{22} = X_{22} - X_{32} g_{23}/g_{33}$$

$$\chi_{66} = X_{66}$$

and for the components of EDC strain matrices $[\Phi]$ and $[\varphi]$

$$\varphi_{11} = \Phi_{11} + g_{13}(\Phi_{33}g_{13} - 2\Phi_{13}g_{33})/g_{33}^2$$

$$\varphi_{12} = \Phi_{12} + (\Phi_{33}g_{13}g_{23} - \Phi_{13}g_{23}g_{33} - \Phi_{23}g_{13}g_{33})/g_{33}^2$$

$$\varphi_{22} = \Phi_{22} + g_{23}(\Phi_{33}g_{23} - 2\Phi_{23}g_{33})/g_{33}^2$$

$$\varphi_{66} = \Phi_{66}$$

(4.95)

Let us represent the relationships between stresses and strains – Hooke's law for plane stress state [Equations (4.51) and (4.52)] in the matrix form with the help of the stiffness matrix $[G]$ or the compliance matrix $[S]$ (see Section 4.1.1)

$$\{\sigma\} = [G]\{\epsilon\}$$

$$\{\epsilon\} = [S]\{\sigma\}$$

(4.96)

Stiffness and compliance matrices are of the forms

$$[G] = \begin{bmatrix} g_{11} & g_{12} & 0 \\ g_{12} & g_{22} & 0 \\ 0 & 0 & g_{66} \end{bmatrix} = \frac{1}{1-\gamma} \begin{bmatrix} E_1 & E_1\nu_{21} & 0 \\ E_2\nu_{12} & E_2 & 0 \\ 0 & 0 & G_{12}(1-\gamma) \end{bmatrix}$$

$$[S] = \begin{bmatrix} s_{11} & s_{12} & 0 \\ s_{12} & s_{22} & 0 \\ 0 & 0 & s_{66} \end{bmatrix} = \begin{bmatrix} 1/E_1 & -\nu_{12}/E_1 & 0 \\ -\nu_{21}/E_2 & 1/E_2 & 0 \\ 0 & 0 & 1/G_{12} \end{bmatrix}$$

(4.97)

Here E_1, E_2, and G_{12} are the Young's moduli along x_1- and x_2-axes and the shear modulus in the (x_1,x_2) plane; ν_{12} and ν_{21} are the Poisson's ratios; $\gamma = \nu_{12}\nu_{21}$ and $E_1\nu_{21} = E_2\nu_{12}$.

Substituting Equation (4.96) in Equation (4.93) and comparing their right parts, one can derive relationships that are analogous to Equation (4.58) but are written in the matrix form:

$$[\psi] = [S][\chi] = [S][\varphi][S]$$

$$[\varphi] = [\chi][G] = [G][\psi][G]$$

(4.98)

$$[\chi] = [G][\psi] = [\varphi][S] = [\chi^*]^T$$

Note that the matrices $[\psi]$ and $[\varphi]$ are symmetric ones, and mixed matrices $[\chi]$ and $[\chi^*]$ are not symmetric.

4.7 COORDINATE TRANSFORMATION

Let us discuss the transformation of elasto-dissipative characteristic matrices [Equation (4.92)] for the transversely isotropic body, as one changes from the (x_1,x_2) coordinate system to any arbitrarily oriented (x_1',x_2') coordinate system. The new coordinate system is turned around the x_3-axis through an angle α with respect to the (x_1,x_2)-axes. The amplitude value of the elastic energy W in a loading cycle is defined by Equation (4.14).

Let us designate the values of matrix components in the (x_1',x_2') coordinates by the primed symbols. To derive the relationships for the matrices $[\psi']$, $[\varphi']$, and $[\chi']$ in the new coordinate system (x_1',x_2'), it is sufficient to exchange stresses and strains in Equation (4.93) for their expressions in the (x_1',x_2')-axes [Equations (4.39) and (4.40)].

The amplitude value of the elastic energy and energy losses in a loading cycle are scalar values and do not depend on the coordinate transformation. Substituting Equations (4.39) and (4.40) into Equations (4.14) and (4.93), it is easy to obtain the transformation rules for corresponding matrices.

Comparison of transformation rules [Equations (4.93) and (4.14)] under the coordinate axes turn shows that the EDC stress matrix, $[\psi]$, transforms in the same way as the compliance matrix, $[S]$; the EDC strain matrix, $[\varphi]$ transforms like the stiffness matrix, $[G]$, viz.,

$$[\psi'] = [T_2][\psi][T_2]^{\mathrm{T}}$$

$$[\varphi'] = [T_1][\varphi][T_1]^{\mathrm{T}}$$

(4.99)

The following transformation formulas are true for the mixed matrix $[\chi']$:

$$[\chi'] = [T_1][\chi][T_2]^{\mathrm{T}} \qquad (4.100)$$

Multiplying the matrices in Equations (4.99) and (4.100), one can derive the relationships between the components of the matrices $[\psi']$, $[\varphi']$, and $[\chi']$ in the (x_1',x_2') coordinate system and the components in the (x_1,x_2) coordinate system:

$$\psi_{11}' = c^4\,\psi_{11} + s^4\,\psi_{22} + (2\psi_{12} + \psi_{66})\,s^2c^2$$

$$\psi_{12}' = (\psi_{11} + \psi_{22} - \psi_{66})\,s^2c^2 + (s^4 + c^4)\,\psi_{12}$$

$$\psi'_{16} = [2\,c^2\,\psi_{11} - 2\,s^2\,\psi_{22} + (2\psi_{12} + \psi_{66})(s^2 - c^2)]\,sc$$

$$\psi'_{22} = s^4\,\psi_{11} + c^4\,\psi_{22} + (2\psi_{12} + \psi_{66})\,s^2c^2 \tag{4.101}$$

$$\psi'_{26} = [2\,s^2\,\psi_{11} - 2\,c^2\,\psi_{22} - (2\psi_{12} + \psi_{66})(s^2 - c^2)]\,sc$$

$$\psi'_{66} = (4\psi_{11} - 8\psi_{12} + 4\psi_{22})\,s^2c^2 + (s^2 - c^2)^2\,\psi_{66}$$

$$\chi'_{11} = c^4\,\chi_{11} + (\chi_{12} + \chi_{21} + 2\chi_{66})\,s^2c^2 + s^4\,\chi_{22}$$

$$\chi'_{12} = c^4\,\chi_{12} + (\chi_{11} + \chi_{22} - 2\chi_{66})\,s^2c^2 + s^4\,\chi_{21}$$

$$\chi'_{16} = 2[c^2\,(\chi_{11} - \chi_{12} - \chi_{66}) - s^2\,(\chi_{22} - \chi_{21} - \chi_{66})]\,sc$$

$$\chi'_{21} = (\chi_{11} + \chi_{22} - 2\chi_{66})\,s^2c^2 + s^4\,\chi_{12} + c^4\,\chi_{21}$$

$$\chi'_{22} = c^4\,\chi_{22} + (\chi_{12} + \chi_{21} + 2\chi_{66})\,s^2c^2 + s^4\,\chi_{11} \tag{4.102}$$

$$\chi'_{26} = 2[s^2\,(\chi_{11} - \chi_{12} - \chi_{66}) - c^2\,(\chi_{22} - \chi_{21} - \chi_{66})]\,sc$$

$$\chi'_{61} = [c^2\,(\chi_{11} - \chi_{21} - \chi_{66}) - s^2\,(\chi_{22} - \chi_{12} - \chi_{66})]\,sc$$

$$\chi'_{62} = [s^2\,(\chi_{11} - \chi_{21} - \chi_{66}) - c^2\,(\chi_{22} - \chi_{12} - \chi_{66})]\,sc$$

$$\chi'_{66} = 2(\chi_{11} + \chi_{22} - \chi_{12} - \chi_{21})\,s^2c^2 + (c^2 - s^2)^2\,\chi_{66}$$

$$\varphi'_{11} = c^4\,\varphi_{11} + s^4\,\varphi_{22} + 2(\varphi_{12} + 2\varphi_{66})\,s^2c^2$$

$$\varphi'_{12} = (\varphi_{11} + \varphi_{22} - 4\varphi_{66})\,s^2c^2 + (s^4 + c^4)\,\varphi_{12}$$

$$\varphi'_{16} = [c^2\,\varphi_{11} - s^2\,\varphi_{22} + (\varphi_{12} + 2\varphi_{66})(s^2 - c^2)]\,sc$$

$$\varphi'_{22} = s^4\,\varphi_{11} + c^4\,\varphi_{22} + 2(\varphi_{12} + 2\varphi_{66})\,s^2c^2 \tag{4.103}$$

$$\varphi'_{26} = [s^2\,\varphi_{11} - c^2\,\varphi_{22} - (\varphi_{12} + 2\varphi_{66})(s^2 - c^2)]\,sc$$

$$\varphi'_{66} = (\varphi_{11} - 2\varphi_{12} + \varphi_{22})\,s^2c^2 + (s^2 - c^2)^2\,\varphi_{66}$$

The relationships between some components of the matrices $[\psi']$, $[\varphi']$,

$[\chi']$ and the angle of coordinate axes turn (α) are shown in Figures 4.2−4.6 for the carbon fiber reinforced plastic (CFRP) HMS/DX 210.

Anisotropic material behavior is known to depend on the orientation of the axes of loading. This dependence may serve as an indicator of the material's anisotropy. As related to dissipative properties, the dependence of the dissipation factor under uniaxial loading along coordinate axes and under ideal shear on the angle of coordinate axes turn may be considered as such an indicator. It is desirable to obtain these dependencies in experimental investigations.

The expression for dissipation factors is defined by a ratio of energy losses [Equation (4.93)] to the amplitude of the elastic energy [Equation (4.14)] in a complete loading cycle. If dissipation factors under uniaxial cyclic loading along the x'_1,x'_2-axes are designated as ψ'_1,ψ'_2 and under ideal shear in (x'_1 and x'_2)-axes as ψ'_6, then

$$\psi'_1 = \frac{\psi'_{11}}{s'_{11}}, \quad \psi'_2 = \frac{\psi'_{22}}{s'_{22}}, \quad \psi'_6 = \frac{\psi'_{66}}{s'_{66}} \qquad (4.104)$$

where ψ'_{11}, ψ'_{22}, and ψ'_{66} are the components of the EDC stress matrix; s'_{11}, s'_{22}, and s'_{66} are the components of the compliance matrix. Let us express Equation (4.104) in terms of engineering elastic and dissipation constants

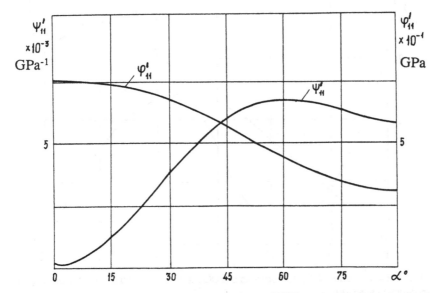

FIGURE 4.2. Values of the components, φ'_{11} and ψ'_{11}, of EDC matrices versus an angle of coordinate axes turn, α.

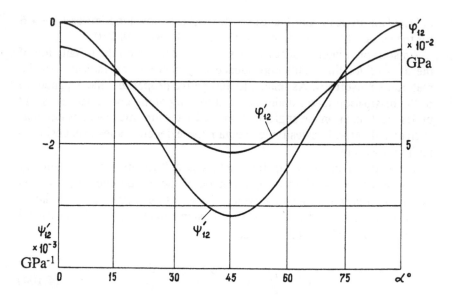

FIGURE 4.3. Values of the components, ψ'_{12} and φ'_{12}, versus an angle of coordinate axes turn, α.

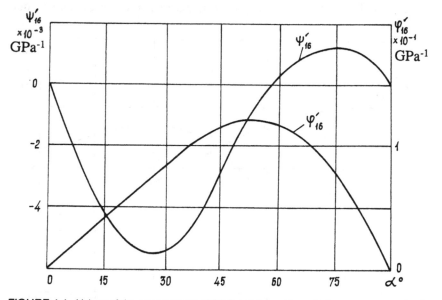

FIGURE 4.4. Values of the components of EDC matrices, ψ'_{16} and φ'_{16}, versus an angle of coordinate axes turn, α.

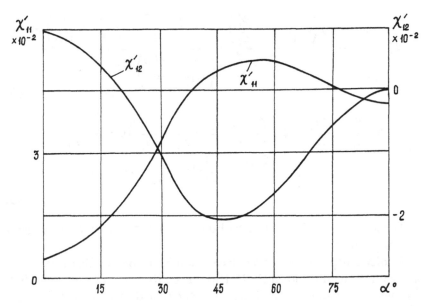

FIGURE 4.5. Values of the components, χ'_{12} and χ'_{11}, versus an angle of coordinate axes turn, α.

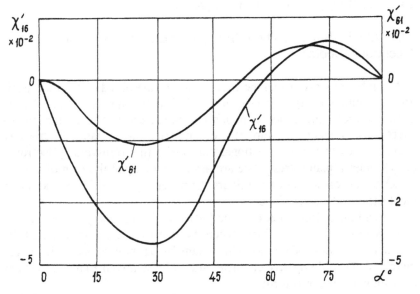

FIGURE 4.6. Values of the components of EDC matrices, χ'_{16} and χ'_{61}, versus an angle of coordinate axes turn, α.

$$\psi_1' = \frac{(c^2 - s^2)\left(c^2 \dfrac{\psi_1^*}{E_1} - s^2 \dfrac{\psi_2^*}{E_2}\right) + \psi_{1/2}^*\left(\dfrac{1 - 2\nu_{12}}{E_1} + \dfrac{1}{E_2} + \dfrac{1}{G_{12}}\right)s^2c^2}{\dfrac{c^4}{E_1} + \dfrac{s^4}{E_2} + \left(\dfrac{1}{G_{12}} - \dfrac{2\nu_{12}}{E_1}\right)s^2c^2}$$

$$(4.105)$$

$$\psi_6' = \frac{4\dfrac{2\psi_1^*}{E_1} + \dfrac{2\psi_1^*}{E_2} - \psi_{1/2}^*\left(\dfrac{1 - 2\nu_{12}}{E_1} + \dfrac{1}{E_2} + \dfrac{1}{G_{12}}\right)s^2c^2 + \dfrac{\psi_6^*}{G_{12}}(s^2 - c^2)^2}{4\left[\dfrac{1 + 2\nu_{12}}{E_1} + \dfrac{1}{E_2}\right]s^2c^2 + \dfrac{1}{G_{12}}(s^2 - c^2)^2}$$

$$(4.106)$$

The formula for ψ_2' is defined by Equation (4.105), provided the angle α is changed for $(\pi/2 - \alpha)$.

Predicted dependences of the dissipation factor on the fiber orientation angle for the unidirectional CFRP HMS/DX 210, HMS/DX 209, HTS/DX 210 and for unidirectional GFRP are represented in Figures 4.7–4.10. Calculations were performed with Equation (4.105). Experimental data belong to the authors of the References [2,52].

4.8 SOME ADDITIONAL COMMENTS

4.8.1 On the Superposition Principle of Energy Losses in Anisotropic Bodies

A number of works using the energy method are based on the superposition principle of energy losses [2,4,44,52,54]. Essentially, the problem is as follows: total energy losses under a complex stress state are the sum of energy losses under the simplest (i.e., uniaxial) stress states. In this way energy losses of the unidirectional composite (the transversely isotropic body) under a plane stress state are assumed to equal the sum of energy losses under loading along the fibers, transversely to the fibers and under ideal shear in the layer plane.

The authors of References [2] and [44] sum the lost energy under simple stress states in the mixed form, when the energy is calculated as half the product of stresses and strains. This means by our notations that the mixed matrix of elasto-dissipative characteristic $[\chi]$ (the mixed EDC matrix) has the following diagonal form:

$$[\chi] = \begin{bmatrix} \chi_{11} & 0 & 0 \\ 0 & \chi_{22} & 0 \\ 0 & 0 & \chi_{66} \end{bmatrix} = \begin{bmatrix} \psi_1^* & 0 & 0 \\ 0 & \psi_2^* & 0 \\ 0 & 0 & \psi_6^* \end{bmatrix}$$

$$(4.107)$$

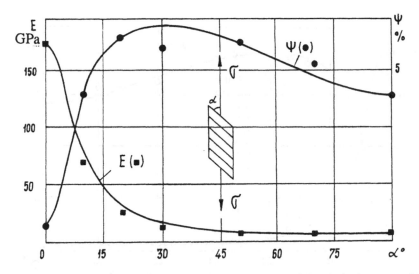

FIGURE 4.7. Unidirectional CFRP HMS/DX 210 experimental data (points) compared to predicted diagrams (full lines) of dissipation factor, $\psi = \psi_1^t$, and Young's modulus, E, versus fiber orientation angle, α. Experimental data are taken from Reference [52].

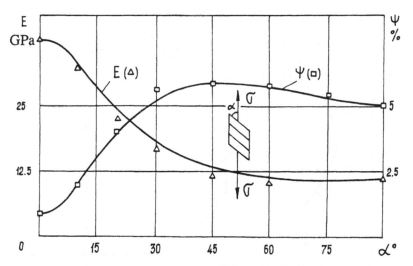

FIGURE 4.8. Unidirectional glass fiber reinforced plastic GLASS/DX 210 experimental data (points) compared to predicted diagrams (full lines) of dissipation factor, $\psi = \psi_1^t$, and Young's modulus, E, versus fiber orientation angle, α. Experimental data are taken from Reference [52].

85

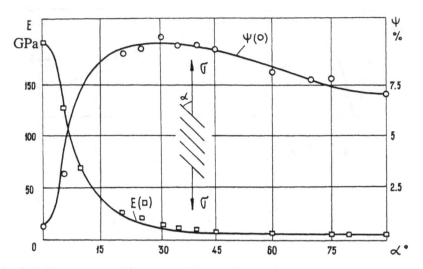

FIGURE 4.9. Unidirectional CFRP HMS/DX 209 experimental data (points) compared to predicted diagrams (full lines) of dissipation factor, ψ, and Young's modulus, E, under uniaxial loading versus fiber orientation angle, α. Experimental data are taken from Reference [2].

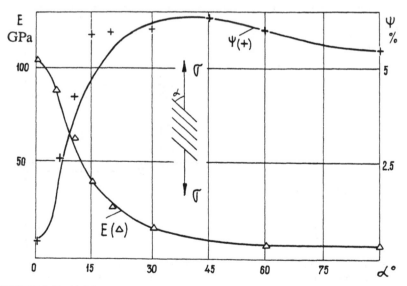

FIGURE 4.10. Unidirectional CFRP HMS/DX 210 experimental data (points) compared to predicted diagrams (full lines) of dissipation factor, ψ, and Young's modulus, E, under uniaxial loading versus fiber orientation angle, α. Experimental data are taken from Reference [2].

It is evident from Equation (4.92) that, following more general energetic Equations (4.53)–(4.55), the transversely isotropic body under plane stress state has an asymmetric EDC matrix of the form [see Equation (4.92)]:

$$[\chi] = \begin{bmatrix} \chi_{11} & \chi_{12} & 0 \\ \chi_{21} & \chi_{22} & 0 \\ 0 & 0 & \chi_{66} \end{bmatrix}$$

$$= \frac{1}{1 - \nu_{12}\nu_{21}} \begin{bmatrix} (1 - \nu_{12}\nu_{21}) & \nu_{12}(\psi_2^* - \psi_1^*) & 0 \\ 0 & -\nu_{12}\nu_{21}(\psi_1^* + \psi_2^*) & 0 \\ 0 & 0 & (1 - \nu_{12}\nu_{21})\psi_6^* \end{bmatrix}$$

$$(4.108)$$

The matrices of Equations (4.107) and (4.108) coincide only if $\nu_{12} = 0$. Naturally, the application of Equations (4.107) and (4.108) gives close results for stress states which are close to simple uniaxial states. The greatest differences between them can take place for biaxial stress states.

Perhaps methodical defects that follow from the superposition principle are more significant. For example, the equality between energy losses under uniaxial stress state ($\sigma_1 \neq 0$, $\sigma_2 = \tau_{12} = 0$) and uniaxial strain state ($\epsilon_1 \neq 0$, $\epsilon_2 = \gamma_{12} = 0$) follows from Equations (4.107) and (4.93). This is physically not very convincing.

A change from the mixed EDC matrix $[\chi]$ [Equation (4.107)] to EDC stress $[\psi]$ or strain $[\varphi]$ matrices [according to Equations (4.98)] results in asymmetrical matrices $[\psi]$ and $[\varphi]$. Asymmetry of these matrices is quite unnatural in completely symmetrical main Equations (4.93).

4.8.2 On the Character of Logarithmic Decrements of Vibrations

Dissipation engineering constants, ψ_i^*, relate to corresponding logarithmic decrements of vibrations [see Equation (2.13)]; these latter are defined experimentally [see Equation (2.4)]. The intuitively realistic and convenient assumption was introduced in References [30] and [69]. The authors believed that logarithmic decrements of the anisotropic body formed a symmetrical tensor of rank 4. This allowed them to formulate corresponding transformation rules for coordinate axes turn.

It is evident from Equations (4.53) and (4.54) that logarithmic decrements do not exhibit tensor features, but more general objects [Equation (4.58)] composed of elastic and dissipation parameters are the tenxors.

This note is likely to be important, as the idea about the tensor nature of vibration decrements has spread into reference books [9], and is still being used in some up-to-date publications [60].

CHAPTER 5

Dissipative Response of Unidirectional Fiber Reinforced Composites: Structural Method of Approach

Elastic and dissipative behavior of composite materials depends on their internal structure and on the properties of their components (a fiber and a matrix). The character of components bond is also important. The problem of predicting effective (mean) composite characteristics on the basis of the characteristics of its constituent components is the subject of composite structural mechanics. This is a large, important, and developing branch of composite mechanics. A considerable number of methods to solve structural problems of composite mechanics were proposed, see for example References [7,16,19,31,32,45,65,85]. The methods can be divided rather conditionally into two groups. The first group can be called the group of "exact" methods that use the solutions of the theory of elasticity or cumbersome calculation procedures (for example, the finite element method). The second group is the group of approximate methods that are based on the methods of approach for the strength of materials.

In spite of obvious advantages of "exact" methods, they are not widespread in practice. At least two reasons can be given for this, to say nothing of the obvious increase of time-consuming analysis.

First, it is difficult to consider a lot of structural features for real composites. Some of these features include complexity and random character of composite internal geometry, the presence of intermediate chemical compounds at the fiber-matrix interface, the presence of delaminations and other defects at the interface, the presence of pores in the matrix, etc.

The second reason relates to the first one. It is difficult to give an impartial assessment of the increase in accuracy that more complex analysis methods give. As a rule, this accuracy falls into the experimental data scattering range. It should be remembered that the data scatter is rather large for composites, which decreases considerably the interest of engineers in seeking such greater accuracy.

The subject of the present chapter is to discuss the structural problem as applied to the dissipative behavior of unidirectional composites. The main goal is not to get reliable quantitative results, but to reveal the principal features of modeling composite dissipative response based on fiber and matrix properties. This chapter also seeks to analyze the correlation between the structural and the phenomenological methods of approach.

The results that are suitable for a clear analysis can be obtained only with the use of fairly simple structural models. The desire for simplicity is the dominant feature of the present chapter. It is clear that there are practically limitless possibilities to refine the results described.

5.1 COMPOSITES OF TRANSVERSELY ISOTROPIC FIBERS AND ISOTROPIC MATRIX: THREE-DIMENSIONAL STRESS STATE

Let us consider a unidirectional material under a three-dimensional (3-D) stress state. The x_1-axis of the Cartesian coordinate system is directed along the fibers (see Figure 5.1), and the axes x_2 and x_3 are perpendicular to the fibers. When determining energy losses in the symmetrical cycle of harmonic loading, let us base our analysis on the following assumptions:

(1) The material, the fibers, and the matrix are linear elastic and uniform; the fibers are transversely isotropic, and the matrix is isotropic.

(2) There are no discontinuities in the displacement field at the fiber-matrix interface, and energy losses in a unit volume of the composite, ΔW, are equal to the sum of energy losses in the fiber, ΔW_f, and in the matrix, ΔW_m, considering their volume fractions

$$\Delta W = \xi \Delta W_f + (1 - \xi) \Delta W_m \qquad (5.1)$$

where ξ is the fiber volume fraction in the composite.

(3) Dissipative properties of composite components are described by five dissipation constants for transversely isotropic fibers and two dissipation constants for the isotropic matrix. Dissipation factors do not depend on the stress amplitude but can depend on temperature and load frequency.

(4) The simplest hypotheses are accepted to describe the stress-strain state of the composite; corresponding components of stress and strain tensors are expressed as follows:

$$\sigma^0_{11} = \xi\sigma^f_{11} + (1 - \xi)\sigma^m_{11}$$

$$\sigma^f_{22} = \sigma^0_{22}, \quad \sigma^m_{22} = \sigma^0_{22}$$

$$\sigma^f_{33} = \sigma^0_{33}, \quad \sigma^m_{33} = \sigma^0_{33}$$

$$\sigma^f_{23} = \sigma^0_{23}, \quad \sigma^m_{23} = \sigma^0_{23}$$

$$\sigma^f_{13} = \sigma^0_{13}, \quad \sigma^m_{13} = \sigma^0_{13}$$

$$\sigma^f_{12} = \sigma^0_{12}, \quad \sigma^m_{12} = \sigma^0_{12}$$

(5.2)

$$\epsilon^f_{11} = \epsilon^0_{11}, \quad \epsilon^m_{11} = \epsilon^0_{11}$$

$$\epsilon^0_{22} = \xi\epsilon^f_{22} + (1 - \xi)\epsilon^m_{22}$$

$$\epsilon^0_{33} = \xi\epsilon^f_{33} + (1 - \xi)\epsilon^m_{33}$$

$$\epsilon^0_{23} = \xi\epsilon^f_{23} + (1 - \xi)\epsilon^m_{23}$$

$$\epsilon^0_{13} = \xi\epsilon^f_{13} + (1 - \xi)\epsilon^m_{13}$$

$$\epsilon^0_{12} = \xi\epsilon^f_{12} + (1 - \xi)\epsilon^m_{12}$$

(5.3)

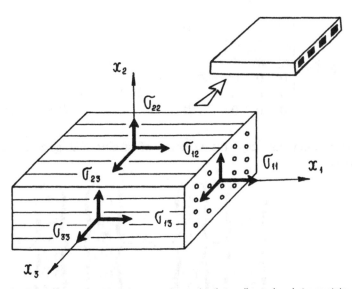

FIGURE 5.1. Unidirectional composite under three-dimensional stress state.

Here σ_{ij}^0, ϵ_{ij}^0 are mean stresses and strains in the composite with respect to the natural axes of anisotropy, σ_{ij}^f, ϵ_{ij}^f, σ_{ij}^m, ϵ_{ij}^m are the components of stress and strain tensors for the fiber and the matrix in the coordinate system chosen.

The "stresses" and the "strains" as well as the "potential energy" are represented by their amplitude values during the harmonic loading cycle. The stresses and the strains in the fiber and in the matrix are related to each other by Hooke's law:

$$\{\epsilon^f\} = [S^f]\{\sigma^f\}$$

$$\{\epsilon^m\} = [S^m]\{\sigma^m\}$$

(5.4)

where compliance matrices for the transversely isotropic fiber and the isotropic matrix are of the form:

$$[S^f] = \begin{bmatrix} s_{11}^f & s_{12}^f & s_{12}^f & 0 & 0 & 0 \\ s_{12}^f & s_{22}^f & s_{23}^f & 0 & 0 & 0 \\ s_{12}^f & s_{23}^f & s_{22}^f & 0 & 0 & 0 \\ 0 & 0 & 0 & 2(s_{22}^f - s_{23}^f) & 0 & 0 \\ 0 & 0 & 0 & 0 & s_{66}^f & 0 \\ 0 & 0 & 0 & 0 & 0 & s_{66}^f \end{bmatrix}$$

(5.5)

$$[S^m] = \begin{bmatrix} s_{11}^m & s_{12}^m & s_{12}^m & 0 & 0 & 0 \\ s_{12}^m & s_{11}^m & s_{12}^m & 0 & 0 & 0 \\ s_{12}^m & s_{12}^m & s_{11}^m & 0 & 0 & 0 \\ 0 & 0 & 0 & 2(s_{11}^m - s_{12}^m) & 0 & 0 \\ 0 & 0 & 0 & 0 & 2(s_{11}^m - s_{12}^m) & 0 \\ 0 & 0 & 0 & 0 & 0 & 2(s_{11}^m - s_{12}^m) \end{bmatrix}$$

The column matrices (the vectors) of strains and stresses have the components:

$$\{\epsilon^f\} = \begin{Bmatrix} \epsilon_{11}^f \\ \epsilon_{22}^f \\ \epsilon_{33}^f \\ 2\epsilon_{23}^f \\ 2\epsilon_{13}^f \\ 2\epsilon_{12}^f \end{Bmatrix} = \begin{Bmatrix} \epsilon_{11}^f \\ \epsilon_{22}^f \\ \epsilon_{33}^f \\ \gamma_{23}^f \\ \gamma_{13}^f \\ \gamma_{12}^f \end{Bmatrix}, \quad \{\epsilon^m\} = \begin{Bmatrix} \epsilon_{11}^m \\ \epsilon_{22}^m \\ \epsilon_{33}^m \\ 2\epsilon_{23}^m \\ 2\epsilon_{13}^m \\ 2\epsilon_{12}^m \end{Bmatrix} = \begin{Bmatrix} \epsilon_{11}^m \\ \epsilon_{22}^m \\ \epsilon_{33}^m \\ \gamma_{23}^m \\ \gamma_{13}^m \\ \gamma_{12}^m \end{Bmatrix}$$

$$\{\sigma^f\} = \begin{Bmatrix} \sigma^f_{11} \\ \sigma^f_{22} \\ \sigma^f_{33} \\ \sigma^f_{23} \\ \sigma^f_{13} \\ \sigma^f_{12} \end{Bmatrix} = \begin{Bmatrix} \sigma^f_{11} \\ \sigma^f_{22} \\ \sigma^f_{33} \\ \tau^f_{23} \\ \tau^f_{13} \\ \tau^f_{12} \end{Bmatrix} \quad \{\sigma^m\} = \begin{Bmatrix} \sigma^m_{11} \\ \sigma^m_{22} \\ \sigma^m_{33} \\ \sigma^m_{23} \\ \sigma^m_{13} \\ \sigma^m_{12} \end{Bmatrix} = \begin{Bmatrix} \sigma^m_{11} \\ \sigma^m_{22} \\ \sigma^m_{33} \\ \tau^m_{23} \\ \tau^m_{13} \\ \tau^m_{12} \end{Bmatrix}$$

here γ^f_{ij}, γ^m_{ij} are shear strains in the corresponding planes for the fiber and the matrix.

Energy losses in the fiber and in the matrix are defined by the relationships given in Chapter 4 for transversely isotropic and isotropic bodies in terms of the stresses

$$\Delta W_f = \frac{1}{2}\{\sigma^f\}^T[\psi^f]\{\sigma^f\}$$

$$\Delta W_m = \frac{1}{2}\{\sigma^m\}^T[\psi^m]\{\sigma^m\}$$

(5.6)

Here the matrix $[\psi^f]$ has the form of Equation (4.88) and the matrix $[\psi^m]$ is the matrix for the isotropic body.

To solve the structural problem on composite dissipative characteristics means to derive the expression of energy losses for the composite as a function of mean stresses. The coefficients in the expression must be determined in terms of the dissipative and elastic parameters of the fiber and the matrix. On the basis of Equations (5.1) and (5.6), it is necessary to know the expressions for the stresses in the fiber and in the matrix in terms of mean composite stresses. Equation (5.2) gives required expressions in the obvious form, except for stress components σ^f_{11}, σ^m_{11}. The equations for stress components selected from Equations (5.2) and (5.3) are as follows:

$$\sigma^0_{11} = \xi\sigma^f_{11} + (1 - \xi)\sigma^m_{11}$$

$$\epsilon^f_{11} = \epsilon^f_{11}$$

(5.7)

Using Hooke's law [Equation (5.4)] and the condition that normal stresses in the fiber along x_2- and x_3-axes and the stresses in the matrix are equal to composite mean stresses, Equation (5.7) is transformed:

$$\sigma^0_{11} = \xi\sigma^f_{11} + (1 - \xi)\sigma^m_{11}$$

$$(s^m_{12} - s^f_{12})(\sigma^0_{22} + \sigma^0_{33}) = s^f_{11}\sigma^f_{11} - s^f_{11}\sigma^f_{11}$$

(5.8)

Solving this set of equations for fiber and matrix stresses, one can derive:

$$\sigma^f_{11} = h^f_{11}\sigma_{11} + h^f_{12}(\sigma_{22} + \sigma_{33})$$

$$\sigma^m_{11} = h^m_{11}\sigma_{11} + h^m_{12}(\sigma_{22} + \sigma_{33})$$

(5.9)

where the coefficients

$$h^f_{11} = \frac{s^m_{11}}{(1-\xi)s^f_{11} + \xi s^m_{11}}; \quad h^f_{12} = \frac{(1-\xi)(s^m_{12} - s^f_{12})}{(1-\xi)s^f_{11} + \xi s^m_{11}}$$

$$h^m_{11} = \frac{s^f_{11}}{(1-\xi)s^f_{11} + \xi s^m_{11}}; \quad h^m_{12} = \frac{\xi(s^f_{12} - s^m_{12})}{(1-\xi)s^f_{11} + \xi s^m_{11}}$$

(5.10)

The required relationship, considering Equations (5.2) and (5.9), will then take the form:

$$\{\sigma^f\} = [H^f]\{\sigma^0\}$$

$$\{\sigma^m\} = [H^m]\{\sigma^0\}$$

(5.11)

Here the matrix $\{\sigma^0\}$ is the matrix of composite mean stresses, and the matrices $[H^f]$ and $[H^m]$ are as follows:

$$[H^f] = \begin{bmatrix} h^f_{11} & h^f_{12} & h^f_{12} & 0 & 0 & 0 \\ 0 & 1 & 0 & 0 & 0 & 0 \\ 0 & 0 & 1 & 0 & 0 & 0 \\ 0 & 0 & 0 & 1 & 0 & 0 \\ 0 & 0 & 0 & 0 & 1 & 0 \\ 0 & 0 & 0 & 0 & 0 & 1 \end{bmatrix}, \quad [H^m] = \begin{bmatrix} h^m_{11} & h^m_{12} & h^m_{12} & 0 & 0 & 0 \\ 0 & 1 & 0 & 0 & 0 & 0 \\ 0 & 0 & 1 & 0 & 0 & 0 \\ 0 & 0 & 0 & 1 & 0 & 0 \\ 0 & 0 & 0 & 0 & 1 & 0 \\ 0 & 0 & 0 & 0 & 0 & 1 \end{bmatrix}$$

Equation (5.1) for energy losses in the composite takes the matrix form

$$\Delta W = \frac{1}{2}\{\sigma^0\}^T[\psi^0]\{\sigma^0\}$$

(5.12)

where $[\psi^0] = \xi[H^f]^T[\psi^f][H^f] + (1-\xi)[H^m]^T[\psi^m][H^m]$ is the matrix of effective dissipative characteristics for a composite of transversely isotropic fibers and an isotropic matrix.

Energy damping in the composite is determined by the ratio between energy losses [Equation (5.12)] and the amplitude value of the potential elastic energy. Elastic energy is determined in terms of mean stresses and the effective compliance matrix $[S^0]$

$$W = \frac{1}{2} \{\sigma^0\}^T [S^0]\{\sigma^0\} \tag{5.13}$$

The matrix $[S^0]$ can be obtained from Equations (5.3) that relate mean composite strains to the strains in the fiber and the matrix with consideration of Hooke's law for composite constituent components and the composite as a whole. These relationships [Equation (5.3)] can be written in the matrix form

$$\{\epsilon^0\} = [D^f]\{\epsilon^f\} + [D^m]\{\epsilon^m\} \tag{5.14}$$

where

$$[D^f] = \begin{bmatrix} 1 & 0 & 0 & 0 & 0 & 0 \\ 0 & \xi & 0 & 0 & 0 & 0 \\ 0 & 0 & \xi & 0 & 0 & 0 \\ 0 & 0 & 0 & \xi & 0 & 0 \\ 0 & 0 & 0 & 0 & \xi & 0 \\ 0 & 0 & 0 & 0 & 0 & \xi \end{bmatrix}$$

$$[D^m] = \begin{bmatrix} 0 & 0 & 0 & 0 & 0 \\ 0 & (1-\xi) & 0 & 0 & 0 \\ 0 & 0 & (1-\xi) & 0 & 0 \\ 0 & 0 & 0 & (1-\xi) & 0 \\ 0 & 0 & 0 & 0 & 0 \\ 0 & 0 & 0 & 0 & (1-\xi) \end{bmatrix}$$

Composite mean strains are expressed in terms of mean stresses according to the law

$$\{\epsilon^0\} = [S^0]\{\sigma^0\} \tag{5.15}$$

Substituting the expressions from Equations (5.14), (5.4), and (5.11) into Equation (5.15), one will obtain the relationship for the matrix of effective compliances

$$[S^0] = [D^f][S^f][H^f] + [D^m][S^m][H^m] \tag{5.16}$$

Here the components of the matrix $[S^0]$ are determined in terms of the components of compliance matrices for composite constituents and fiber volume fraction.

Equations (5.12), (5.13), and (5.16) define composite effective dissipative and elastic characteristics in terms of the characteristics of the isotropic matrix and transversely isotropic fibers.

5.2 PLANE STRESS STATE FOR A COMPOSITE OF TRANSVERSELY ISOTROPIC FIBERS AND ISOTROPIC MATRIX

The order of the matrices and their form in Equations (5.4)–(5.6) and (5.11)–(5.16) become simpler in the case of the plane stress state [in the (x_1,x_2) plane]. It is enough to consider three-component column matrices of stresses and strains

$$\{\sigma^f\} = \begin{Bmatrix} \sigma^f_{11} \\ \sigma^f_{22} \\ \tau^f_{12} \end{Bmatrix}, \quad \{\sigma^m\} = \begin{Bmatrix} \sigma^m_{11} \\ \sigma^m_{22} \\ \tau^m_{12} \end{Bmatrix}, \quad \{\epsilon^f\} = \begin{Bmatrix} \epsilon^f_{11} \\ \epsilon^f_{22} \\ \gamma^f_{12} \end{Bmatrix}, \quad \{\epsilon^m\} = \begin{Bmatrix} \epsilon^m_{11} \\ \epsilon^m_{22} \\ \gamma^m_{12} \end{Bmatrix} \quad (5.17)$$

The matrices $[H^f]$ and $[H^m]$ in Equation (5.11), fiber and matrix compliance matrices $[S^f]$ and $[S^m]$ will be of the form:

$$[H^f] = \begin{bmatrix} h^f_{11} & h^f_{12} & 0 \\ 0 & 1 & 0 \\ 0 & 0 & 1 \end{bmatrix}, \quad [H^m] = \begin{bmatrix} h^m_{11} & h^m_{12} & 0 \\ 0 & 1 & 0 \\ 0 & 0 & 1 \end{bmatrix}$$

$$(5.18)$$

$$[S^f] = \begin{bmatrix} s^f_{11} & s^f_{12} & 0 \\ s^f_{12} & s^f_{22} & 0 \\ 0 & 0 & s^f_{66} \end{bmatrix}, \quad [S^m] = \begin{bmatrix} s^m_{11} & s^m_{12} & 0 \\ s^m_{12} & s^m_{11} & 0 \\ 0 & 0 & 2(s^m_{11} - s^m_{12}) \end{bmatrix}$$

Equation (5.10) is true for the components h^f_{ij} and h^m_{ij} of the matrices $[H^f]$ and $[H^m]$. The matrices of elasto-dissipative characteristic $[\psi^f][\psi^m]$ are of the form:

$$[\psi^f] = \begin{bmatrix} \psi^f_{12} & \psi^f_{12} & 0 \\ \psi^f_{12} & \psi^f_{22} & 0 \\ 0 & 0 & \psi^f_{66} \end{bmatrix}, \quad [\psi^m] = \begin{bmatrix} \psi^m_{12} & \psi^m_{12} & 0 \\ \psi^m_{12} & \psi^m_{22} & 0 \\ 0 & 0 & 2(\psi^m_{11} - \psi^m_{12}) \end{bmatrix} \quad (5.19)$$

and the matrices for the strain components $[D^f]$ and $[D^m]$ take the following form:

$$[D^f] = \begin{bmatrix} 1 & 0 & 0 \\ 0 & \xi & 0 \\ 0 & 0 & \xi \end{bmatrix}, \quad [D^m] = \begin{bmatrix} 0 & 0 & 0 \\ 0 & (1-\xi) & 0 \\ 0 & 0 & (1-\xi) \end{bmatrix} \quad (5.20)$$

The matrices with the dimension 3×3 for composite elasto-dissipative characteristic and composite compliances are described by Equations (5.12) and (5.16).

The dissipative behavior of the unidirectional composite under a plane stress state is characterized by four independent constants.

5.3 COMPOSITE OF ISOTROPIC FIBERS IN ISOTROPIC MATRIX: THREE-DIMENSIONAL STRESS STATE

Let us now examine the case where the isotropy of the properties is a feature not only of the matrix, but of the fibers as well. The assumptions of analysis formulated in Section 5.1 of the present chapter still apply, except for the third of them which is changed to the following form:

- Specific energy losses in composite constituents are equal to the sum of energy losses due to shear and volume strains

$$\Delta W_f = \Delta W_f' + \Delta W_f'', \; \Delta W_m = \Delta W_m' + \Delta W_m'' \tag{5.21}$$

  where a single prime corresponds to the form change energy and a double prime corresponds to the volume change energy.
- Specific energy losses in composite components during the loading cycle are proportional to amplitude values of the form and volume change energies

$$\Delta W_f' = \psi_f' W_f', \; \Delta W_f'' = \psi_f'' W_f''$$
$$\Delta W_m' = \psi_m' W_m', \; \Delta W_m'' = \psi_m'' W_m'' \tag{5.22}$$

Here ψ_f', ψ_m' are dissipation factors for the fiber and the matrix, which correspond to the form change energy for the fiber W_f' and the matrix W_m'; ψ_f'' and ψ_m'' are dissipation factors for the fiber and the matrix, which correspond to the volume change energy for the fiber W_f'' and the matrix W_m''; dissipation factors of composite constituents do not depend on the stress amplitude, but in general can depend on temperature and load frequency.

Let us consider the expression for the specific energies of composite constituents [Equation (5.21)] in terms of the stresses in the fiber and in the matrix. The form change energy is determined as [48]:

$$W_x' = \sigma_x \mathfrak{z}_x / 2, \; \sigma_x = \sqrt{s_{ij}^x s_{ij}^x}, \; \mathfrak{z}_x = \sqrt{\mathfrak{z}_{ij}^x \mathfrak{z}_{ij}^x} \tag{5.23}$$

where the stress deviator s_{ij}^x and the strain deviator $\mathfrak{z}_{ij}^x$ are expressed in terms of stress tensors σ_{ij}^x and strain tensors ϑ_{ij}^x and the Kronecker delta δ_{ij} (the superscript x should be changed to the superscripts f and m for the fiber and the matrix respectively).

$$s_{ij}^x = \sigma_{ij}^x - \delta_{ij}\sigma_{ll}/3, \quad \vartheta_{ij}^x = \epsilon_{ij}^x - \delta_{ij}\epsilon_{ll}/3 \tag{5.24}$$

The volume change energy is expressed in the form:

$$W_x'' = \frac{1}{2}\frac{\sigma_{ii}^x}{3}\,\epsilon_{ij}^x \tag{5.25}$$

Considering the relationship between the deviators and first invariants of stress and strain tensors $\sigma_x = 2G_x\,\vartheta_x$, $\sigma_{ii}^x = 3K_x\epsilon_{ii}^x$, using Equations (5.21), (5.22), (5.23) and (5.25), one can derive the relation for complete specific energy losses in the fiber and in the matrix

$$\Delta W_x = \psi_x'\frac{\sigma_x^2}{4G_x} + \psi_x''\frac{(\sigma_{ii}^x)^2}{18K_x} \tag{5.26}$$

Here G_f and G_m are shear moduli, while K_f and K_m are bulk moduli for the fiber and the matrix respectively. These moduli are related to Young's moduli E_f and E_m and Poisson's ratios ν_f and ν_m in the form

$$G_x = \frac{E_x}{2(1 + \nu_x)}, \quad K_x = \frac{E_x}{3(1 - 2\nu_x)} \tag{5.27}$$

Equation (5.26) with the help of Equation (5.27) takes the following form:

$$\Delta W_x = \frac{1}{2E_x}\left[\psi_x'(1 + \nu_x)\sigma_x^2 + \frac{\psi_x''}{3}(1 - 2\nu_x)(\sigma_{ii}^x)^2\right]$$

or in the detailed form:

$$\Delta W_x = \frac{1}{2E_x}\left\{\frac{2\psi_x'(1 + \nu_x)}{3}\left[\sigma_{11}^{x2} + \sigma_{22}^{x2} + \sigma_{33}^{x2} - \sigma_{11}^x\sigma_{22}^x - \sigma_{11}^x\sigma_{33}^x - \sigma_{22}^x\sigma_{33}^x\right.\right.$$

$$+ 3(\sigma_{12}^{x2} + \sigma_{23}^{x2})\bigg] + \frac{\psi_x''}{3}(1 - 2\nu_x)\left[\sigma_{11}^{x2} + \sigma_{22}^{x2} + \sigma_{33}^{x2}\right.$$

$$\left.\left.+ 2(\sigma_{11}^x\sigma_{22}^x + \sigma_{11}^x\sigma_{33}^x + \sigma_{22}^x\sigma_{33}^x)\right]\right\} \tag{5.28}$$

Let us note that the tensors σ_{ij}^x are amplitude values of stresses in composite constituents in one complete harmonic loading cycle.

Equation (5.11) for isotropic fiber and matrix are expressed in the form [7]:

$$\sigma_{11}^f = \frac{E_f}{E_1}\,\sigma_{11}^0 + (1 - \xi)\,\frac{k}{E_1}\,(\sigma_{22}^0 + \sigma_{33}^0)$$

$$\sigma_{11}^m = \frac{E_m}{E_1}\,\sigma_{11}^0 + \xi\,\frac{k}{E_1}\,(\sigma_{22}^0 + \sigma_{33}^0)$$

$$\sigma_{22}^f = \sigma_{22}^0, \quad \sigma_{33}^f = \sigma_{33}^0$$

$$\sigma_{22}^m = \sigma_{22}^0, \quad \sigma_{33}^m = \sigma_{33}^0 \tag{5.29}$$

$$\tau_{12}^f = \tau_{12}^0, \quad \tau_{13}^f = \tau_{13}^0, \quad \tau_{23}^f = \tau_{23}^0$$

$$\tau_{12}^m = \tau_{12}^0, \quad \tau_{13}^m = \tau_{13}^0, \quad \tau_{23}^m = \tau_{23}^0$$

where σ_{ij}^0 are mean amplitude stress values in the unidirectional composite, and E_1 is the Young's modulus of the composite along the fiber direction, which is determined according to the rule of mixtures

$$E_1 = \xi E_f + (1 + \xi)E_m \tag{5.30}$$

and k is a parameter of the form $k = \nu_m E - \nu_f E_m$.

Considering Equations (5.1), (5.28) and (5.29), the expression for specific dissipated energy of the unidirectional composite in a complete loading cycle takes the form

$$\Delta W = \frac{1}{2}\,[A\sigma_{11}^{02} + B(\sigma_{22}^{02} + \sigma_{33}^{02}) + C(\sigma_{12}^{02} + \sigma_{13}^{02} + \sigma_{23}^{02})$$

$$+ D(\sigma_{11}^0\sigma_{22}^0 + \sigma_{11}^0\sigma_{33}^0) + F\sigma_{22}^0\sigma_{33}^0] \tag{5.31}$$

where the coefficients A, B, C, D, and F are equal to

$$A = \frac{1}{3E_1^2}\,\{\xi E_f\,[2\psi_f'(1 + \nu_f) + \psi_f''(1 - 2\nu_f)]$$

$$+ (1 - \xi)E_m[2\psi_m'(1 + \nu_m) + \psi_m''(1 - 2\nu_m)]\}$$

$$B = \frac{1}{3}\left\{\frac{\xi}{E_f}[2\psi_f'(1 + \nu_f) + \psi_f''(1 - 2\nu_f)] + \frac{(1 - \xi)}{E_m}[2\psi_m'(1 + \nu_m)\right.$$

$$\left. + \psi_m''(1 - 2\nu_m)] + \frac{2k\xi(1 - \xi)}{E_1}\left(\frac{\psi_f' - \psi_f''}{E_f} - \frac{\psi_m' - \psi_m''}{E_m}\right)\right\}$$

$$C = 2\left[\frac{\xi\psi_f'(1 + \nu_f)}{E_f} + \frac{(1 - \xi)\psi_m'(1 + \nu_m)}{E_m}\right] \qquad (5.32)$$

$$D = \frac{2}{3E_1}\left\{\frac{\xi(1 - \xi)k}{E_1}(2\psi_m' + \psi_m'' - 2\psi_f' - \psi_f'') + \xi[\psi_f''(1 - 2\nu_f)\right.$$

$$\left. - \psi_f'(1 + \nu_f)] + (1 - \xi)[\psi_m''(1 - 2\nu_m) - \psi_m'(1 + \nu_m)]\right\}$$

$$F = \frac{2\xi}{3E_f}[2(1 - \xi)\frac{k}{E_1}(\psi_f' - \psi_f'') - \psi_f'(1 + \nu_f) + \psi_f''(1 - 2\nu_f)]$$

$$+ \frac{2(1 - \xi)}{3E_m}[2\xi\frac{k}{E_1}(\psi_m' - \psi_m'') + \psi_m''(1 - 2\nu_m) - \psi_m'(1 + \nu_m)]$$

In Equation (5.32) the terms with the order of the square of Poisson's ratio are ignored, as they are of small magnitude compared to the unity.

Relationships in Equations (5.31) and (5.32) make it possible to determine energy losses in the unidirectional composite under an arbitrary 3-D stress state. They should be treated as approximate relationships, as energy losses in real composites can be concerned not only with absorption of energy by the fibers and the matrix, but also with the presence of intermediate chemical compounds, cracks, pores, etc. It is difficult to estimate their effect a priori.

Let us represent Equation (5.31) in the matrix form in terms of EDC stress matrix

$$\Delta W = \frac{1}{2}\{\sigma^0\}^T[\Psi]\{\sigma^0\} \qquad (5.33)$$

where the components of [Ψ] matrix are as follows

$$[\Psi] = \begin{bmatrix} A & D/2 & D/2 & 0 & 0 & 0 \\ & B & F/2 & 0 & 0 & 0 \\ & & B & 0 & 0 & 0 \\ & & & C & 0 & 0 \\ \text{sim.} & & & & C & 0 \\ & & & & & C \end{bmatrix}$$

and $\{\sigma^0\}$ is the column matrix of stresses.

The unidirectional composite can be considered as a uniform transversely isotropic elastic body. It was shown in Chapter 4 that the matrix [Ψ] for such a body has the form of Equation (4.88), i.e., energy dissipation under 3-D stress state is described with the help of five components of the EDC stress matrix. Due to the existence of the isotropic plane, the component ψ_{44} is expressed in terms of the components ψ_{22} and ψ_{23} as follows: $\psi_{44} = 2(\psi_{22} - \psi_{23})$. It is easy to ensure that in a structural approach for determination of matrix [Ψ] components the relationship mentioned is retained with new notations:

$$C = 2B - F \tag{5.34}$$

Therefore, structural analysis shows the mechanism for modeling the dissipative behavior of the unidirectional composite under an arbitrary stress state.

Except for Equation (5.34), one more additional relationship between the components of the [Ψ] matrix takes place:

$$\psi_{44} = \psi_{66} \tag{5.35}$$

The relationship mentioned follows from some assumptions that are characteristic for the chosen model of the material. In fact, it was assumed in Equation (5.29), that

$$\tau_{23}^m = \tau_{23}^f = \tau_{23}^0$$

and

$$\tau_{12}^m = \tau_{12}^f = \tau_{12}^0$$

These result for isotropic fibers in the appearance of the additional Equation (5.35).

In this case the dissipative behavior of the material is defined by four constants, for example, A, B, C, and D.

The values of these coefficients can be determined in the experiments of cyclic loading. Let us consider the unidirectional composite under longitudinal tension. Energy losses during the loading cycle and the amplitude value of the elastic energy depend on the stress amplitude

$$\Delta W = \frac{1}{2} A \sigma_{11}^2, \quad W = \frac{1}{2} \frac{\sigma_{11}^2}{E_1}$$

The dissipation factor in this case

$$\psi_1^* = \Delta W / W = A E_1 = \psi_{11} E_1 \qquad (5.36)$$

The expressions for the dissipation factor, ψ_2^*, when loading the composite in the direction transverse to the fiber direction (along the x_2-axis) and the dissipation factor, ψ_6^*, when loading the composite in shear in the (x_1, x_2) plane, will be derived in an analogous way:

$$\psi_2^* = B E_2 = \psi_{22} E_2, \quad \psi_6^* = C G_{12} = \psi_{66} G_{12} \qquad (5.37)$$

Dissipation factors that conform to uniaxial loading along the x_3-axis, ψ_3^*, as well as to the loading at the $45°$ angle to the x_1-axis in the (x_1, x_2) plane, $\psi_{1/2}^*$, and the factors that conform to shear in the (x_2, x_3) and (x_1, x_3) planes, ψ_4^* and ψ_5^*, take the following form:

$$\psi_3^* = B E_3, \quad \psi_{1/2}^* = \frac{A + B + C + D}{\dfrac{1 - 2\nu_{12}}{E_1} + \dfrac{1}{E_2} + \dfrac{1}{G_{12}}} \qquad (5.38)$$

$$\psi_4^* = C G_{23}, \quad \psi_5^* = C G_{13}$$

Universally adopted nomenclature for the transversely isotropic body is used here — E_1, E_2, and E_3 are Young's moduli; G_{12}, G_{13}, and G_{23} are shear moduli; ν_{12} is the Poisson's ratio, in this case

$$E_2 = E_3, \quad G_{12} = G_{13} \qquad (5.39)$$

Three equations for dissipation factors of the unidirectional material follow from formulas $(5.37) - (5.39)$

$$\psi_2^* = \psi_3^*, \quad \psi_5^* = \psi_6^*, \quad \psi_6^* / G_{12} = \psi_4^* / G_{23} \qquad (5.40)$$

A set of coefficients ψ_1^*, ψ_2^*, ψ_6^*, and $\psi_{1/2}^*$ can be considered as a system of four independent engineering constants that are determined from direct experiments under simple types of stress states. These coefficients govern energy dissipation in cyclic loading of the unidirectional composite under an arbitrary 3-D stress state.

The components of the effective compliance matrix are defined in terms of engineering elastic constants [7]:

$$[S] = \begin{bmatrix} 1/E_1 & -\nu_{21}/E_2 & -\nu_{12}/E_2 & 0 & 0 & 0 \\ & 1/E_2 & -\nu_{12}/E_2 & 0 & 0 & 0 \\ & & 1/E_2 & 0 & 0 & 0 \\ & \text{sim.} & & 1/G_{23} & 0 & 0 \\ & & & & 1/G_{13} & 0 \\ & & & & & 1/G_{12} \end{bmatrix} \quad (5.41)$$

where engineering constants E_1 and E_2 are Young's moduli under tension along the fibers and transverse to the fibers, G_{23} is the shear modulus in the isotropic plane, G_{13} and G_{12} ($G_{13} = G_{12}$) are shear moduli in the planes that are perpendicular to the isotropic plane, and ν_{12} and ν_{21} are Poisson's ratios. For the case of isotropy of fiber and matrix properties [7] the expressions for engineering elastic constants are of the form:

$$E_1 = \xi E_f + (1 - \xi)E_m$$

$$E_2 = \frac{E_f E_m}{\xi E_m + (1 - \xi)E_f}$$

$$\quad (5.42)$$

$$G_{13} = G_{12} = \frac{G_f G_m}{\xi G_m + (1 - \xi)G_f}$$

$$\nu_{12} = \xi \nu_f + (1 - \xi)\nu_m, \quad \nu_{21} = \frac{\nu_{12} E_2}{E_1}$$

Here E_f, E_m, G_f, G_m, ν_f, and ν_m are Young's moduli, shear moduli, and Poisson's ratios of the fiber and the matrix.

It should be remembered that Equations (5.42) are true if Equations (5.2) and (5.3) are met, and that the terms on the order of the square of Poisson's ratio are neglected since they are of small magnitude compared to the unity.

Let us consider the specific case of relationships for the constants A, B, C, and D [Equation (5.32)]. The constant F is related to the constants B and C by Equation (5.34).

Now suppose that the fiber does not have the ability to dissipate energy, i.e., $\psi_f' = \psi_f'' = 0$. Then it follows from Equation (5.32)

$$A = \frac{(1 - \xi)E_m}{E_1^2} [2\psi_m' (1 + \nu_m) + \psi_m'' (1 - 2\nu_m)]$$

$$B = \frac{(1 - \xi)}{3E_m} [2\psi_m' (1 + \nu_m) + \psi_m''(1 - 2\nu_m) - \frac{2k\xi}{E_1} (\psi_m' - \psi_m'')]$$

$$C = \frac{2(1 - \xi)\psi_m'(1 + \nu_m)}{E_m} \tag{5.43}$$

$$D = \frac{2(1 - \xi)}{3E_1} \left[\frac{\xi k}{E_1} (2\psi_m' + \psi_m'') - \psi_m'(1 + \nu_m) + \psi_m''(1 - 2\nu_m) \right]$$

When substituting Equation (5.30) for the parameters k and E_1 into the expression for the coefficient B and using the rule of mixture for composite Poisson's ratio

$$\nu_{12} = \xi\nu_f + (1 - \xi)\nu_m \tag{5.44}$$

the expression for B becomes simpler.

Let us represent the expressions for the coefficients A, C, and D in another form, using shear modulus, G_m, and bulk modulus, K_m, of the matrix, Equation (5.27). Then, the coefficients will be written as:

$$A = \frac{(1 - \xi)E_m^2}{3E_1^2} \left[\frac{\psi_m'}{G_m} + \frac{\psi_m''}{3K_m} \right]$$

$$B = \frac{(1 - \xi)}{3} \left[\frac{2\psi_m' + \psi_m''}{E_m} + \frac{2\nu_{12}}{E_1} (\psi_m' - \psi_m'') \right]$$

$$C = \frac{(1 - \xi)\psi_m'}{G_m} \tag{5.45}$$

$$D = \frac{2(1 - \xi)E_m}{3E_1} \left[\frac{\xi k}{E_1 E_m} (2\psi_m' + \psi_m'') - \frac{\psi_m'}{2G_m} + \frac{\psi_m''}{3K_m} \right]$$

Let us assume that there is no energy dissipation in the fiber and in the matrix during volume deformation. Only the coefficient, ψ_m', differs from zero among the four dissipation factors ψ_f', ψ_f'', ψ_m', and ψ_m''. Then Equation (5.45) is transformed as follows:

$$A = (1 - \xi)\,\psi_m'\,\frac{E_m^2}{3E_1^2 G_m}$$

$$B = (1 - \xi)\psi_m'\frac{2}{3}\left(\frac{1}{E_m} + \frac{\nu_{12}}{E_1} \right)$$

$$C = (1 - \xi)\psi_m'\frac{1}{G_m} \tag{5.46}$$

$$D = (1 - \xi)\psi'_m \frac{2}{3} \frac{\xi(E_f \nu_m - E_m \nu_f) - \nu_{12}E_m}{E_1^2} - \frac{1}{E_1}$$

All the coefficients A, B, C, and D in Equation (5.46) are proportional to the product $(1 - \xi)\psi'_m$, but absolute values of these constants may differ considerably. In fact, real composites are made of rigid fibers and a compliant matrix, i.e., $E_f \gg E_m$, $E_f \gg G_m$; matrix shear and Young's moduli are of the same order, fiber Young's modulus and composite longitudinal modulus are of the same order: $G_m \sim E_m$, $E_f \sim E_1$.

Based on the conditions mentioned, estimating the absolute values of coefficients in Equation (5.46), one can get

$$B \sim C, \quad B \gg D, \quad C \gg D, \quad D \gg A \tag{5.47}$$

Therefore in the sum [Equation (5.31)] that defines energy losses of the unidirectional composite under an arbitrary 3-D stress state, the coefficient for σ_{11}^2 has the least absolute value. The coefficients B, C, and F prevail here [see Equation (5.34)], while the coefficient D has an intermediate value. One must take care that the estimation given is true only for absolute values of the coefficients A, B, C, D, and F. The contribution of each element to the total sum of energy losses depends on the composite stress state. In practice, the composite is designed so that the fibers carry the main load. For example, normal stresses along the fiber direction are usually much greater than the stresses in the transverse direction, and the addendum of $A\sigma_{11}^2$, $B\sigma_{22}^2$, . . . may be of the same order.

It is possible to use Equations (5.46) to calculate all the coefficients A, B, C, D, and F from one experimental value of a composite dissipation factor. In doing so, it is sufficient to know the values of the following elastic constants: matrix Young's modulus and shear modulus, E_m and G_m, longitudinal elastic modulus of the composite, E_1, Poisson's ratio, ν_{12}, and fiber volume fraction, ξ. The rest of the elastic parameters are calculated through the rule of mixture.

In the final analysis, the application of Equations (5.46) to a specific composite is determined by comparison of experimental data and theoretical results.

For many real isotropic bodies, energy losses in one complete loading cycle are characterized not by the amplitudes of form and/or volume changes, but by the amplitude of complete specific elastic energy of the body in a loading cycle (see, for example, References [48,57,61]). This means that the dissipative behavior of the body is determined only by one parameter, the energy dissipation factor, ψ. Such a problem formulation is the special case of Equation (5.28), when dissipation factors relating to form and volume changes are equal, $\psi = \psi' = \psi''$. Total specific energy losses in the body, ΔW, are proportional to the amplitude of elastic energy, W

$$\Delta W = \psi W$$

From this it should be supposed in Equation (5.32) that $\psi_f = \psi_f' = \psi_f''$ $\psi_m = \psi_m' = \psi_m''$, then

$$A = \frac{1}{E_1^2} [\psi_f E_f \xi + \psi_m E_m (1 - \xi)]$$

$$B = \frac{\xi \psi_f}{E_f} + \frac{(1 - \xi) \psi_m}{E_m}$$

$$C = \frac{\xi \psi_f}{G_f} + \frac{(1 - \xi) \psi_m}{G_m}$$

$$D = \frac{2}{E_1^2} [\xi \nu_f + (1 - \xi) \nu_m][\xi \psi_f E_f + (1 - \xi) \psi_m E_m]$$

(5.48)

Here Equation (5.34) is true for B, C, and F, and the coefficient D is the dependent one. Considering the expression for Poisson's ratio ν_{12}, one can get

$$D = -2\nu_{12}A \qquad (5.49)$$

Existence of the simple relationship between the coefficients A and D, shown in Equation (5.49) enables one to reduce the number of independent elasto-dissipative parameters of the unidirectional composite under arbitrary 3-D stress state to three parameters. The symmetric elasto-dissipative stress matrix $[\Psi]$, Equation (5.33), with consideration of Equation (5.34) takes the form

$$[\Psi] = \begin{bmatrix} A & -\nu_{12}A & -\nu_{12}A & 0 & 0 & 0 \\ & B & B - C/2 & 0 & 0 & 0 \\ & & B & 0 & 0 & 0 \\ & \text{sim.} & & C & 0 & 0 \\ & & & & C & 0 \\ & & & & & C \end{bmatrix} \qquad (5.50)$$

Figures 5.2–5.4 show the predicted relationships of relative energy dissipation and elastic moduli versus fiber volume fraction in unidirectional composites under simple types of loading. Input data for calculation are tabulated in Table 5.1. The curves with number 1 correspond to the most general case of energy dissipation both in the matrix and in the fiber, Equations (5.32). The curves with number 2 conform to the model with one energy dissipation factor in the matrix and in the fiber, Equations (5.48) (a single-component model). It is believed that for the curves with number 3, energy dissipation takes place in the matrix due only to the form change.

Attempts to test Equations (5.32) experimentally run into technical difficulties, as it is necessary to determine a large number of dissipation factors and elastic constants for the matrix and the fiber. That is why experimental plots for dissipation factors in the reinforcing direction versus fiber volume fraction [92] are compared with Equations (5.48). The values of the dissipation factors, as well as the Young's moduli of the fiber and the matrix were calculated from experimental data when $\xi = 0$ and $\xi = 0.5$, using the rule of mixture [Equation (5.30)]. Theoretical and experimental data in Figures 5.5 and 5.6 are in good agreement.

The phenomenological method of approach makes it possible to determine the values of all coefficients in Equations (5.48) and (5.50) from the experiments in uniaxial cyclic loading along and transverse to the fibers and in ideal shear in the (x_1, x_2)-axes. As a result the corresponding dissipation factors ψ_1^*, ψ_2^*, and ψ_6^* (engineering constants) can be obtained. The coefficients A, B, and C equal

$$A = \frac{\psi_1^*}{E_1}, \quad B = \frac{\psi_2^*}{E_2}, \quad C = \frac{\psi_6^*}{G_{12}} \qquad (5.51)$$

The components of the matrix [Ψ] [Equation (5.50)] are expressed in terms of dissipation factors – engineering dissipation constants – and in terms of engineering elastic constants – Young's moduli E_1, E_2, shear modulus G_{12}, and Poisson's ratio ν_{12}:

TABLE 5.1. Dissipation Constants for Three Calculation Variants.

No.	ψ_f', %	ψ_f'', %	ψ_m', %	ψ_m'', %
1	2	1	8	4
2	2	2	8	8
3	0	0	8	0

Elastic constants are the following: $E_f = 74$ GPa, $\nu_f = 0.2$, $E_m = 3$ GPa, $\nu_m = 0.35$.

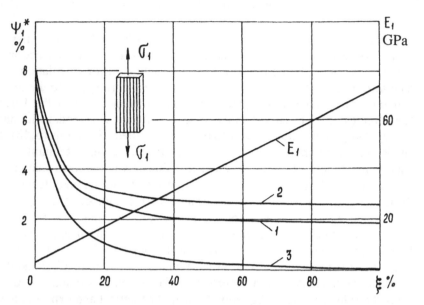

FIGURE 5.2. Predicted relationships of dissipation factor, ψ_1^*, and Young's modulus, E_1, during loading along the fibers versus fiber volume fraction. For initial data, see Table 5.1.

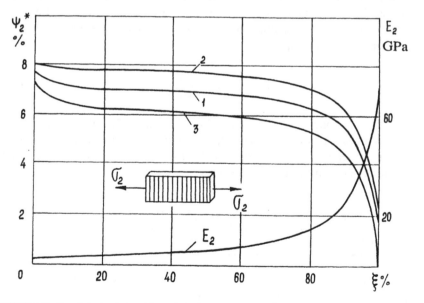

FIGURE 5.3. Predicted relationships of dissipation factor, ψ_2^*, and Young's modulus, E_2, during loading transverse to the fibers versus fiber volume fraction. For initial data, see Table 5.1.

108

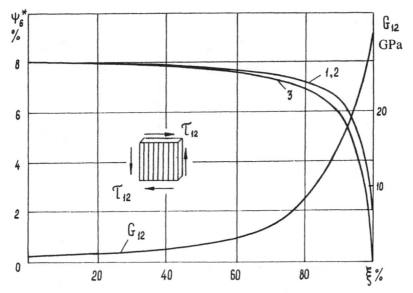

FIGURE 5.4. Predicted relationships of dissipation factor, ψ_6^*, and shear modulus, G_{12}, during ideal shear versus fiber volume fraction, ξ. For initial data, see Table 5.1.

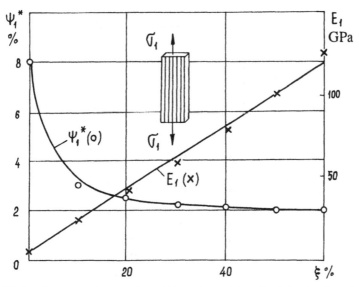

FIGURE 5.5. Dissipation factor, ψ_1^*, and Young's modulus, E_1, during loading along the fibers versus relative fiber volume fraction, ξ. Full lines—calculation results, points—experimental data [92]. Initial data for calculation: $E_f = 194$ GPa, $\nu_f = 0.3$, $\psi_f = 1.8\%$, $E_m = 6$ GPa, $\nu_m = 0.3$, $\psi_m = 8.0\%$.

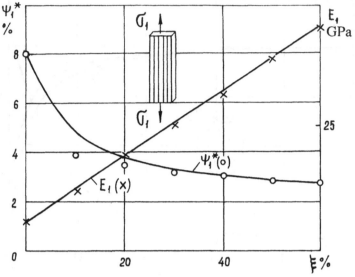

FIGURE 5.6. Dissipation factor, ψ_1^*, and Young's modulus, E_1, during loading along the fibers versus relative fiber volume fraction, ξ. Full lines—calculation results, points—experimental data [92]. Initial data for calculation: $E_f = 71$ GPa, $\nu_f = 0.3$, $\psi_f = 2.5\%$, $E_m = 6$ GPa, $\nu_m = 0.3$, $\psi_m = 8.0\%$.

$$
[\Psi] = \begin{bmatrix}
\psi_1^*/E_1 & -\dfrac{\nu_{12}\psi_1^*}{E_1} & -\dfrac{\nu_{12}\psi_1^*}{E_1} & 0 & 0 & 0 \\
 & \psi_2^*/E_2 & \dfrac{\psi_2^*}{E_1} - \dfrac{\psi_6^*}{2G_{12}} & 0 & 0 & 0 \\
 & & \psi_2^*/E_2 & 0 & 0 & 0 \\
 & & & \psi_6^*/G_{12} & 0 & 0 \\
 & \text{sim.} & & & \psi_6^*/G_{12} & 0 \\
 & & & & & \psi_6^*/G_{12}
\end{bmatrix}
$$

(5.52)

5.3.1 Example

Let us calculate the elasto-dissipative matrix, $[\Psi]$, and the compliance matrix, $[S]$, for the unidirectional composite with isotropic constituents, using the single-component model for dissipative characteristics of the fiber and the matrix. Initial data are the following:

$$E_f = 194 \text{ GPa}, \quad \nu_f = 0.3, \quad \psi_f = 0.018$$

$$E_m = 6 \text{ GPa}, \quad \nu_m = 0.3, \quad \psi_m = 0.08, \quad \xi = 0.3$$

To calculate the components of the matrices, let us use Equations (5.42) and (5.48). The components have the following values:

$$\psi_{11} = 0.358 \times 10^{-3} \qquad s_{11} = 0.016$$

$$\psi_{22} = 0.859 \times 10^{-2} \qquad s_{22} = 0.109$$

$$\psi_{12} = 0.107 \times 10^{-3} \qquad s_{12} = 0.24 \times 10^{-2}$$

$$\psi_{66} = 2.444 \times 10^{-1} \qquad s_{66} = 0.309$$

Dissipation factors of the composite under uniaxial loading along the fibers, ψ_1^*, transverse to the fibers, ψ_2^*, and under ideal shear in the (x_1,x_2) plane, ψ_6^*, in accordance with Equations (5.36)−(5.38) are equal to

$$\psi_1^* = 0.022, \quad \psi_2^* = 0.078, \quad \psi_6^* = 0.079$$

5.4 PLANE STRESS STATE FOR A COMPOSITE OF ISOTROPIC FIBERS AND ISOTROPIC MATRIX

Under a plane stress state in the axes (x_1,x_2), when $\sigma_{13}^0 = \sigma_{23}^0 = \sigma_{33}^0 = 0$, energy losses can be represented with the help of the reduced EDC stress matrix [w] and the reduced stress matrix $\{\sigma^0\} = \{\sigma_{11}^0, \sigma_{22}^0, \sigma_{12}^0\}^T$

$$\Delta W = \{\sigma^0\}^T [\psi] \{\sigma^0\}$$

$$[\psi^0] = \begin{bmatrix} \psi_{11} & \psi_{12} & 0 \\ \psi_{12} & \psi_{22} & 0 \\ 0 & 0 & \psi_{66} \end{bmatrix} = \begin{bmatrix} A & D/2 & 0 \\ D/2 & B & 0 \\ 0 & 0 & C \end{bmatrix} \qquad (5.53)$$

The matrix, $[\psi^0]$, is analogous to the corresponding elasto-dissipative stress matrix from Chapter 4 in the number of independent components and the ways of their determination in terms of dissipation factors (engineering constants). It is possible to turn from the matrix, $[\psi^0]$ [Equation (5.53)] to the elasto-dissipative strain matrix $[\varphi]$ and the mixed matrix, $[\chi]$ [Equation (4.98)]. When transforming the coordinate system by rotation of the (x_1, x_2)-axes about the x_3-axis, the components ψ_{ij}, φ_{ij}, and χ_{ij} $(i,j = 1,2,6)$ of the matrices $[\psi]$, $[\varphi]$, and $[\chi]$ are calculated in accordance with the rules of Equations $(4.101) - (4.103)$.

Let us write down specific energy losses in one complete loading cycle for the composite under a plane stress state. This can be done with the help of the reduced elasto-dissipative stress matrix $[\Psi]$, and the stress matrix, $\{\sigma^0\}$, in the following form:

$$\Delta W = \{\sigma^0\}^{\mathrm{T}}[\psi^0]\{\sigma^0\} = \begin{Bmatrix} \sigma_{11}^0 \\ \sigma_{22}^0 \\ \sigma_{12}^0 \end{Bmatrix}^{\mathrm{T}} \begin{bmatrix} \dfrac{\psi_1^*}{E_1} & -\dfrac{\nu_{12}\psi_1^*}{E_1} & 0 \\[3mm] -\dfrac{\nu_{12}\psi_1^*}{E_1} & \dfrac{\psi_2^*}{E_2} & 0 \\[3mm] 0 & 0 & \dfrac{\psi_6^*}{G_{12}} \end{bmatrix} \begin{Bmatrix} \sigma_{11}^0 \\ \sigma_{22}^0 \\ \sigma_{12}^0 \end{Bmatrix} \qquad (5.54)$$

Next let us apply Equation (4.98) to the elasto-dissipative strain matrix $[\varphi]$, and the mixed EDC matrix, $[\chi]$ [Equation (4.97)]. Taking into account the expressions for the components of the stiffness matrix, $[G]$, and the compliance matrix, $[S]$, in terms of engineering elastic constants, one can derive the expressions for the matrices $[\chi]$ and $[\varphi]$

$$[\chi] = \frac{1}{1 - \gamma^2} \begin{bmatrix} (1 - \gamma)\psi_1^* & \nu_{12}(\psi_2^* - \psi_1^*) & 0 \\ 0 & -\gamma\psi_1^* + \psi_2^* & 0 \\ 0 & 0 & (1 - \gamma)\psi_6^* \end{bmatrix}$$

$$(5.55)$$

$$[\varphi] = \frac{1}{(1 - \gamma)^2} \begin{bmatrix} E_1[\psi_1^* - \gamma(\psi_2^* - 2\psi_1^*)] & E_1\nu_{21}(\psi_2^* - \gamma\psi_1^*) & 0 \\ \nu_{12}E_2(\psi_2^* - \gamma\psi_1^*) & E_2(\psi_2^* - \gamma\psi_1^*) & 0 \\ 0 & 0 & (1 - \gamma)^2\psi_6^*G_{12} \end{bmatrix}$$

where $\gamma = \nu_{12}\nu_{21}$. Among the components of each EDC matrix [Equations (5.54) and (5.55)] only three components are independent, since the following relationships hold:

$$\psi_{12} = -\nu_{12}\psi_{11}$$

$$\chi_{12} = \nu_{12}(\chi_{22} - \chi_{11}) \qquad (5.56)$$

$$\varphi_{12} = \nu_{12}\varphi_{22}$$

Note that when changing from a 3-D stress state of the unidirectional composite [Equation (5.52)] to a plane stress state, the number of independent engineering constants (those that describe energy losses in the material) remains the same and is equal to three.

Therefore, the structural method of approach for determining the dissipative response of the unidirectional composite under an arbitrary 3-D stress state via the energy method results in two models for the system of independent dissipation constants. In the first case the specific dissipated energy (EDC matrices) is described by the system of four coefficients A, B, C, and D, according to Equation (5.33). In the second case, energy losses are described by the system of three constants A, B, and C, according to Equation (5.50), provided that Equation (5.49) is true. The first system of dissipation parameters for the unidirectional composite material is termed the four-component model, and the second system is termed the three-component model. The three-component model makes it possible to examine some problems in convenient analytical form and to derive clear geometrical interpretations.

CHAPTER 6

Effective Viscoelastic Characteristics of Fiber Reinforced Composites

One model to describe energy damping during vibrations is the model of the linear viscoelastic body. Expressions that relate stresses and strains in the form of Volter integral relationships are of the most general type [see Equation (4.78)]. When considering stationary vibrations, it is convenient to turn to the complex form of stress and strain representation and to introduce complex moduli, Equations (4.80) and (4.81). Complex moduli characterize energy losses in a complete loading cycle, as in Equation (4.85).

Introduction of complex moduli produces effective methods for solving both the problems of composite mechanics and the problems of stationary vibrations in viscoelastic structural elements. For elastic and viscoelastic materials under harmonic vibrations, it is sufficient to have equations that relate complex σ_{ij}^* and ϵ_{ij}^*, instead of common state equations (the relationships $\sigma_{ij} \sim \epsilon_{ij}$). Complex moduli C_{ijkl}^* must be considered for viscoelastic bodies [see Equation (4.80)].

Thus, to solve problems of stationary harmonic viscoelastic vibrations, it is sufficient to use the solutions of the problems within the theory of elasticity. It is only necessary to exchange the moduli of elasticity for complex moduli of the viscoelastic body and to separate real and imaginary parts in the solutions obtained.

This principle is a special case of the general correspondence principle in the theory of viscoelasticity. Usually, in order to obtain the solution, it is necessary to invert Laplace-Karson transformations (see, for example, Reference [65]). In the present problem formulation it is sufficient to exchange real moduli for complex ones.

General relations for composite complex moduli in terms of complex moduli of constituent materials (a fiber and a matrix) are derived in this chapter. Comparison is also made between two different methods of approach—the energy method and the method of complex moduli.

115

6.1 COMPLEX COMPLIANCES OF THE VISCOELASTIC BODY

When calculating the dissipation factors that correspond to specific cases of the stress-strain state, it is more convenient to operate with the compliance characteristics. Physical relations of the theory of linear viscoelasticity, which are written in terms of the tensor of creep functions $S_{ijkl}(t)$, are of the form

$$\epsilon_{ij}(t) = \int_{-\infty}^{t} S_{ijkl}(t - \tau)d\sigma_{kl}(\tau) \tag{6.1}$$

where t is time, and $\epsilon_{ij}(t)$ and $\sigma_{kl}(t)$ are strain and stress tensors.

Let us introduce complex stresses, σ_{ij}^*, for cyclic harmonic loading with the frequency, ω, and the stress amplitude, σ_{ij}^0

$$\sigma_{ij}^*(t) = \sigma_{ij}^0 e^{-i\omega t} \tag{6.2}$$

Substitution of Equation (6.2) into Equation (6.1) results in the concept of complex compliances of the viscoelastic anisotropic body

$$\epsilon_{ij}^* = S_{ijkl}^* \sigma_{kl}^* \tag{6.3}$$

Here $\epsilon_{ij}^*(t)$ is the tensor of harmonically varying complex strains, S_{ijkl}^* is the tensor of complex compliances that includes the real S_{ijkl}' and the imaginary S_{ijkl}'' parts

$$S_{ijkl}^* = S_{ijkl}' + iS_{ijkl}''$$

$$S_{ijkl}'(\omega) = \omega \int_{0}^{\infty} S_{ijkl}(u) \sin(\omega u) du \tag{6.4}$$

$$S_{ijkl}''(\omega) = \omega \int_{0}^{\infty} S_{ijkl}(u) \cos(\omega u) du$$

The following designations are accepted in Equation (6.3) and further: index "*" denotes a complex value, a prime and a double prime denote real and imaginary parts of the value respectively. Complex compliances as well as complex moduli depend on the load frequency ω and on the temperature. Let us note that complex moduli and compliances [Equation (6.3)] can be interpreted as the images of Laplace-Karson transformations of relaxation and creep functions. These images are known to be mutually inverse [65], which is why $S^*_{ijkl}C^*_{ijkl} = 1$.

Transforming Equation (6.3), one obtains the obvious form of the relationship between the complex strain and time

$$\epsilon^*_{ij} = S'_{ijkl}\sigma^0_{kl}\cos\omega t - S''_{ijkl}\sigma^0_{kl}\sin\omega t + i(S'_{ijkl}\sigma^0_{kl}\sin\omega t - S''_{ijkl}\sigma^0_{kl}\cos\omega t)$$

$$(6.5)$$

Equation (6.5) is represented in the form

$$\epsilon^*_{ij}(t) = \epsilon^0_{ij}e^{i(\omega t+\theta_{ij})} \qquad \text{(no summing!)} \qquad (6.6)$$

where

$$\epsilon^0_{ij} = \sqrt{i(S'_{ijkl}\sigma^0_{kl})^2 + i(S''_{ijkl}\sigma^0_{kl})^2}$$

$$\text{tg }\theta_{ij} = \frac{S''_{ijkl}\sigma^0_{kl}}{S'_{ijkl}\sigma^0_{kl}} \qquad \text{(no summing!)} \qquad (6.7)$$

Here ϵ^0_{ij} is the amplitude of the periodic strain, and θ_{ij} is interpreted as the strain lag angle with relation to stress during stationary vibrations of the viscoelastic body. Sometimes the value tg θ_{ij} is referred to as the loss tangent (loss factor) [19].

Note that the values tg θ_{ij} and ϵ^0_{ij} are not tensors in general, and ϵ^*_{ij} is a tensor. It follows directly from the analysis of right parts of Equation (6.5).

Energy losses ΔW in the unit volume of a body in a complete loading cycle are determined with the help of the integral

$$\Delta W = \oint\sigma_{ij}d\epsilon_{ij} = \oint \text{Re}\sigma^*_{ij}d(\text{Re}\epsilon^*_{ij}) \qquad (6.8)$$

or, considering Equations (6.3) and (6.4)

$$\Delta W = -\pi S''_{ijkl}\sigma^0_{ij}\sigma^0_{kl} \qquad (6.9)$$

The imaginary part of the complex compliances coincides with the elasto-dissipative characteristic stress tensor to an accuracy of a constant multiplier. Recall that the elasto-dissipative characteristic stress (strain) tensor was introduced in Chapter 4.

Dissipation factor ψ is the ratio of energy losses [Equation (6.9)] to the amplitude value of the potential energy W in a complete loading cycle

$$\psi = \frac{\Delta W}{W} = -\frac{2\pi S''_{ijkl}\sigma^0_{ij}\sigma^0_{kl}}{S'_{ijkl}\sigma^0_{ij}\sigma^0_{kl}} \tag{6.10}$$

where

$$W = \frac{1}{2}S'_{ijkl}\sigma^0_{ij}\sigma^0_{kl}$$

Equation (6.10) enables one to calculate energy losses and the dissipation factor under different types of stress states, which are defined by the values of the stress tensor components σ^0_{ij}.

6.2 EFFECTIVE COMPLEX CHARACTERISTICS OF THE UNIDIRECTIONAL COMPOSITE IN TERMS OF FIBER AND MATRIX CHARACTERISTICS

The authors of References [31] and [32] were among the first to investigate composite complex moduli in terms of matrix and filler complex moduli. The authors assumed that energy losses occurred only through the matrix form change. Effective complex moduli for the unidirectional fibrous composite under a plane stress state were derived in Reference [32]. The authors of Reference [32] believed that the fibers were absolutely elastic and that the matrix had a complex shear modulus and a real bulk modulus.

Let us consider a more general case. The matrix and the fiber are assumed to be isotropic, linear viscoelastic, uniform materials. The unidirectional material is under a plane stress state. The x_1-axis is directed along the fibers and the x_2-axis is directed transversely to the fibers (x_1 and x_2 are the coordinate axes of the unidirectional composite). Viscoelastic properties of both the matrix and the fiber under stationary harmonic loading are each determined by two parameters. These are independent complex moduli that are analogous to Young's modulus and shear modulus

$$E^*_m = E'_m + iE''_m, \qquad G^*_m = G'_m + iG''_m$$

$$E^*_f = E'_f + iE''_f, \qquad G^*_f = G'_f + iG''_f \tag{6.11}$$

where the subscript m is for matrix parameters and the subscript f is for fiber parameters.

Complex Young's moduli and shear moduli are determined from the functions of stress relaxation, which correspond to the uniaxial loading and ideal shear of the isotropic matrix and fiber.

Complex Poisson's ratios of the matrix and the fiber are expressed in terms of complex Young's and shear moduli as follows:

$$\nu_x^* = \frac{E_x^*}{2G_x^*} - 1 = \frac{E_x'G_x' - E_x''G_x'' + i(E_x''G_x' + E_x'G_x'')}{2|G_x^*|^2} - 1 \qquad (6.12)$$

Here the subscript x must be changed to m or f for the matrix and the fiber respectively, and $|G_x^*|$ is the absolute value of the complex shear modulus.

It is customary to determine the values of the dissipation factors in the experiments on energy damping. Let us express the dissipation factors in terms of imaginary and real parts of complex moduli. Dissipation factors for the matrix and the fiber in uniaxial loading are designated as ψ_m^+ and ψ_f^+, and in ideal shear are designated as ψ_m' and ψ_f'. Then, using Equation (6.10), one can obtain

$$\psi^+ = -\frac{2\pi S_{1111}''}{S_{1111}'}, \qquad \psi' = -\frac{2\pi S_{1212}''}{S_{1212}'} \qquad (6.13)$$

where the components of the complex compliances are expressed in terms of complex Young's and shear moduli in the form:

$$S_{1111}^* = \frac{1}{E^*} = \frac{E' - iE''}{|E^*|}$$

$$\qquad (6.14)$$

$$S_{1212}^* = \frac{1}{4G^*} = \frac{G' - iG''}{4|G^*|}$$

The indices for the matrix and the fiber are omitted in Equations (6.13) and (6.14).

Finally we have the following relationships:

$$\psi_m^+ = 2\pi\frac{E_m''}{E_m'}, \quad \psi_f^+ = 2\pi\frac{E_f''}{E_f'}, \quad \psi_m' = 2\pi\frac{G_m''}{G_m'}, \quad \psi_f' = 2\pi\frac{G_f''}{G_f'} \qquad (6.15)$$

The dissipative response of the viscoelastic constituents of the composite is defined by four independent dissipation factors.

Further, it is assumed that effective elastic characteristics of the unidirectional composite under plane stress state—Young's moduli along and transverse to the fibers E_1 and E_2, shear modulus G_{12}, and Poisson's ratio ν_{12}—are determined from the following simple formulas:

$$E_1 = \xi E_f + (1 - \xi)E_m, \quad E_2 = \frac{E_f E_m}{\xi E_m + (1 - \xi)E_f}$$

$$G_{12} = \frac{G_f G_m}{\xi G_m + (1 - \xi)G_f}, \quad \nu_{12} = \xi \nu_f + (1 - \xi)\nu_m$$

(6.16)

where ξ is the fiber volume fraction in the material, and the rest of the parameters have universally adopted nomenclature.

Let us now exchange the values of the elastic constants in Equation (6.16) for complex Young's moduli, shear modulus and Poisson's ratio [Equations (6.11) and (6.12)] and select imaginary and real parts. The following relations will then be derived for complex moduli and the complex Poisson's ratio of the composite:

$$E_1^* = E_1' + iE_1'' = [\xi E_f' + (1 - \xi)E_m'] + i[\xi E_f'' + (1 - \xi)E_m'']$$

$$E_2^* = E_2' + iE_2'' = [\xi E_f'|E_m^*|^2 + (1 - \xi)E_m'|E_f^*|^2]/l$$

$$+ i[\xi E_f''|E_m^*|^2 + (1 - \xi)E_m''|E_f^*|^2]/l$$

$$G_{12}^* = G_{12}' + iG_{12}'' = [\xi G_f'|G_m^*|^2 + (1 - \xi)G_m'|G_f^*|^2]/g$$

$$+ i[\xi G_f''|G_m^*|^2 + (1 - \xi)G_m''|G_f^*|^2]/g$$

(6.17)

$$\nu_{12}^* = \nu_{12}' + i\nu_{12}'' = \frac{\xi(E_f' G_f' + E_f'' G_f'')}{2|G_f^*|^2} + \frac{(1 - \xi)(E_m'G_m' + E_m''G_m'')}{2|G_m^*|^2}$$

$$- 1 + i\left[\frac{\xi(E_f' G_f' - E_f' G_f'')}{2|G_f^*|^2} + \frac{(1 - \xi)(E_m''G_m' - E_m'G_m'')}{2|G_m^*|^2}\right]$$

where

$$l = [\xi E_m' + (1 - \xi)E_f']^2 + [\xi E_m'' + (1 - \xi)E_f'']^2$$

$$g = [\xi G_m' + (1 - \xi)G_f']^2 + [\xi G_m'' + (1 - \xi)G_f'']^2$$

Equation (6.17) makes it possible to estimate the effect of the viscoelastic complex parameters of the matrix and the fiber, as well as the fiber volume fraction, on the effective complex characteristics and dissipative properties of the composite. The relationships expressed in Equation (6.17) should be considered estimated ones, due to the inevitable effects of different factors that are not considered in the adopted structural model of composite material.

Reference [32] considers a particular case of Equation (6.17) in which one supposes that $E_f'' = 0$, $G_f'' = 0$ and the imaginary part of the matrix complex bulk modulus equals zero. This means that energy losses in composite constituents occur only through the matrix form change.

Complex characteristics of the viscoelastic body can be represented in the matrix form, as it has been done for elastic characteristics. The expressions for their components in terms of the engineering constants—Young's moduli, shear modulus and Poisson's ratio—remain the same, if one substitutes the values of complex constants [Equation (6.17)] instead of real elastic constants. For example, the matrix of complex compliances $[S*]$ is of the form:

$$[S*] = [S'] + i[S''] = \begin{bmatrix} s_{11}' & s_{12}' & 0 \\ s_{12}' & s_{22}' & 0 \\ 0 & 0 & s_{66}' \end{bmatrix} + i \begin{bmatrix} s_{11}'' & s_{12}'' & 0 \\ s_{12}'' & s_{22}'' & 0 \\ 0 & 0 & s_{66}'' \end{bmatrix}$$

$$(6.18)$$

$$[S*] = \begin{bmatrix} s_{11}^* & s_{12}^* & 0 \\ s_{12}^* & s_{22}^* & 0 \\ 0 & 0 & s_{66}^* \end{bmatrix} = \begin{bmatrix} 1/E_1^* & -\nu_{12}^*/E_1^* & 0 \\ -\nu_{12}^*/E_1^* & 1/E_2^* & 0 \\ 0 & 0 & 1/G_{12}^* \end{bmatrix}$$

Let us represent complex stresses and strains as matrices $\{\sigma*\}$ and $\{\epsilon*\}$, and the amplitude values of stresses as the matrix $\{\sigma^0\}$

$$\{\sigma*\} = \begin{Bmatrix} \sigma_{11}^* \\ \sigma_{22}^* \\ \sigma_{12}^* \end{Bmatrix} = \begin{Bmatrix} \sigma_{11}' \\ \sigma_{22}' \\ \sigma_{12}' \end{Bmatrix} + i \begin{Bmatrix} \sigma_{11}'' \\ \sigma_{22}'' \\ \sigma_{12}'' \end{Bmatrix}, \quad \{\sigma*\} = \{\sigma^0\}e^{-i\omega t}$$

$$(6.19)$$

$$\{\epsilon*\} = \begin{Bmatrix} \epsilon_{11}^* \\ \epsilon_{22}^* \\ \epsilon_{12}^* \end{Bmatrix} = \begin{Bmatrix} \epsilon_{11}' \\ \epsilon_{22}' \\ \epsilon_{12}' \end{Bmatrix} + i \begin{Bmatrix} \epsilon_{11}'' \\ \epsilon_{22}'' \\ \epsilon_{12}'' \end{Bmatrix}, \quad \{\sigma^0\} = \begin{Bmatrix} \sigma_{11}^0 \\ \sigma_{22}^0 \\ \sigma_{12}^0 \end{Bmatrix}$$

Equations (6.3), (6.9), and (6.10) for the plane stress state can be transformed to the matrix form

$$\{\epsilon^*\} = [S^*]\{\sigma^*\}, \quad \Delta W = -\pi\{\sigma^0\}^T[S'']\{\sigma^0\}, \quad W = \frac{1}{2}\{\sigma^0\}^T[S']\{\sigma^0\}$$

(6.20)

Four components of the complex compliance matrix are expressed in terms of complex moduli and complex Poisson's ratio from the formulas:

$$s_{11}^* = \frac{E_1' - iE_1''}{|E_1^*|^2}, \quad s_{12}^*E^* = -\frac{\nu_{12}'E_1' + \nu_{12}''E_1'' + i(\nu_{12}''E_1' - \nu_{12}'E_1'')}{|E_1^*|^2}$$

(6.21)

$$s_{22}^* = \frac{E_2' - iE_2''}{|E_2^*|^2}, \quad s_{66}^* = \frac{G_{12}' - iG_{12}''}{|G_{12}^*|^2}$$

The values of the engineering dissipation parameters of the composite (dissipation factors) are usually defined from direct experiments in cyclic loading of the specimens. The specimens are cut out at different angles to the fiber direction. The compliance matrix $[S^*]$ [Equation (6.18)] includes four independent components. This means that the corresponding four independent dissipation factors describe the dissipative response of the composite.

Let us derive the expressions for the dissipation factors in terms of the imaginary and real parts of complex compliances and complex moduli of the composite, using Equations (6.20) and (6.21). The equations will be written as:

$$\psi_1^* = -2\pi\frac{s_{11}''}{s_{11}'} = 2\pi\frac{E_1''}{E_1'}, \quad \psi_2^* = -2\pi\frac{s_{22}''}{s_{22}'} = 2\pi\frac{E_2''}{E_2'}$$

(6.22)

$$\psi_6^* = -2\pi\frac{s_{66}''}{s_{66}'} = 2\pi\frac{G_{12}''}{G_{12}'}, \quad \psi_{1/2}^* = -2\pi\frac{s_{11}'' + 2s_{12}'' + s_{66}''}{s_{11}' + 2s_{12}' + s_{66}'}$$

Here ψ_1^*, ψ_2^*, and $\psi_{1/2}^*$ are the dissipation factors under uniaxial loading of the composite along the fibers, transverse to the fibers, and at an angle of $45°$ to the fiber direction respectively, while ψ_6^* is the dissipation factor under ideal shear in the plane (x_1,x_2).

The real parts of the components of the complex compliance matrix, Equation (6.21), are the dynamic compliances of the viscoelastic body. For many materials, some composites being among them, these compliances show poor dependence on the load frequency in a rather wide range. That is why it may be assumed that viscoelastic moduli or compliances are equal to "elastic" moduli or compliances of the material. Imaginary parts of the components of the compliance matrix [Equation (6.21)] are expressed in terms of dissipation engineering constants – dissipation factors, if one solves Equation (6.22) for the values s_{11}'', s_{22}'', s_{12}'', and s_{66}''.

6.2.1 Example

Let us calculate the composite complex moduli and dissipation factors from the complex moduli of an isotropic matrix and fibers.

Initial data are the following:

(1) Elastic constants

$$E_f' = 140 \text{ GPa}, \quad v_f' = 0.2, \quad E_m' = 3 \text{ GPa}, \quad v_m' = 0.35$$

(2) Dissipation factors under uniaxial loading and ideal shear
  • the fibers

$$\psi_f^+ = 0.018, \qquad \psi_f' = 0.02$$

  • the matrix

$$\psi_m^+ = 0.076, \qquad \psi_m' = 0.08$$

(3) Fiber volume fraction
$$\xi = 0.3$$

Imaginary parts of the complex moduli for the fibers and the matrix are calculated from Equation (6.22):

$$E_f'' = \frac{\psi_f^+}{2\pi} E_f' = 0.401 \text{ GPa}, \quad G_f'' = \frac{\psi_f'}{2\pi} G_f' = 0.186 \text{ GPa}$$

where

$$G_f' = \frac{E_f'}{2(1 + v_f')} = 58 \text{ GPa}$$

$$E_m'' = \frac{\psi_m^+}{2\pi} E_m' = 0.036 \text{ GPa}, \quad G_m'' = \frac{\psi_m'}{2\pi} G_m' = 0.014 \text{ GPa}$$

where

$$G_m' = \frac{E_m'}{2(1 + \nu_m')} = 1.1 \text{ GPa}$$

Composite complex moduli are calculated from Equation (6.17):

$$E_1^* = 44.1 + i0.145 \text{ GPa}, \qquad E_2^* = 4.3 + i0.051 \text{ GPa}$$

$$G_{12}^* = 1.6 + i0.02 \text{ GPa}, \qquad \nu_{12}^* = 0.30 + i0.72 \times 10^{-3}$$

Dissipation factors of the composite under simple types of stress states are determined on the basis of Equations (6.21) and (6.22):

$$\psi_1^* = 0.021, \quad \psi_2^* = 0.075, \quad \psi_6^* = 0.079, \quad \psi_{1/2}^* = 0.078$$

Figures 6.1–6.3 illustrate predicted variations of real and imaginary parts of complex moduli with fiber volume fraction ξ in the composite. The following data were used as input data for calculating Equation (6.22): $E_f' = 74$ MPa, $\nu_f' = 0.2$, $E_m' = 3$ MPa, $\nu_m' = 0.35$, $\psi_f' = 0.02$, $\psi_f^* = 0.018$, $\psi_m' = 0.08$, $\psi_m^* = 0.076$.

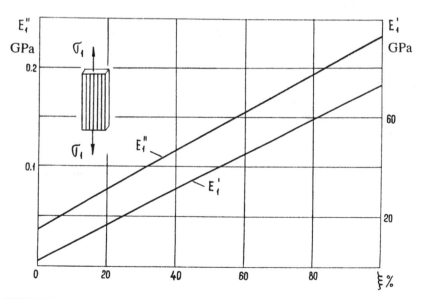

FIGURE 6.1. Predicted relationships between the real (E_1') and the imaginary (E_1'') parts of longitudinal complex Young's modulus and fiber volume fraction, ξ.

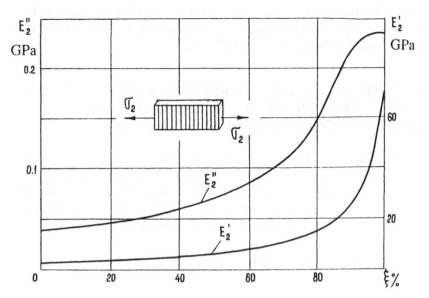

FIGURE 6.2. Predicted relationships between the real (E_2') and the imaginary (E_2'') parts of transverse complex Young's modulus and fiber volume fraction, ξ.

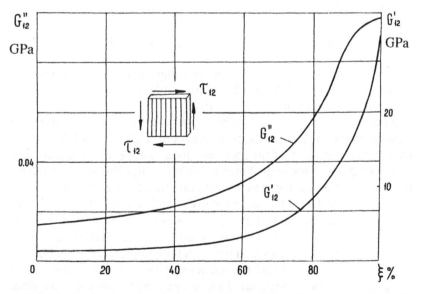

FIGURE 6.3. Predicted relationships between the real (G_{12}') and the imaginary (G_{12}'') parts of complex shear modulus and fiber volume fraction, ξ.

125

To define the imaginary parts of the matrix and fiber complex moduli, Equation (6.22) was used. Note that correspondence relationships for real and imaginary parts are of a "similar" type.

6.3 COMPARISON OF COMPOSITE DISSIPATIVE CHARACTERISTICS DERIVED BY THE METHOD OF COMPLEX MODULI AND THE ENERGY METHOD

The energy method of approach was used in Chapter 5 to determine the coefficients in the relationships for energy lost in a complete loading cycle. The coefficients were expressed in terms of elasto-dissipative characteristic matrices for an isotropic matrix and isotropic fibers. The formula for energy losses under a plane stress state is of the form [Equation (5.43)]:

$$\Delta W = \frac{1}{2}(A\sigma_{11}^2 + D\sigma_{11}\sigma_{22} + B\sigma_{22}^2 + C\sigma_{12}^2) \qquad (6.23)$$

where the coefficients A, B, C, and D relate to dissipation factors, fiber and matrix moduli of elasticity, and fiber volume fraction [Equation (6.13)].

If the method of complex compliances is applied, then one has from Equation (6.20)

$$\Delta W = -\pi(s_{11}''\sigma_{11}^2 + 2s_{12}''\sigma_{11}\sigma_{22} + s_{22}''\sigma_{22}^2 + s_{66}''\sigma_{12}^2) \qquad (6.24)$$

Here the imaginary parts of complex compliances s_{ij} are expressed in terms of complex moduli and the relative volume fraction of composite components by Equations (6.17) and (6.21).

Let us compare results obtained by two different methods. Begin by comparing corresponding coefficients in Equations (6.23) and (6.24), provided that the imaginary parts of the complex moduli for the matrix and the fiber are exchanged with the help of inverted relationships [Equation (6.15)] and their real parts are accepted as the common moduli of elasticity.

Dissipation factors for form change, ψ', and volume change, ψ'', were taken as the dissipation engineering constants of composite isotropic constituents in Equation (5.43). Dissipation factors under uniaxial loading, ψ^+, and ideal shear, ψ', were used in the model of complex moduli in Equation (6.15). It is necessary to find the relationship between the coefficients ψ'' and ψ^+ for an isotropic material. This can be done by two methods, which correspond to the energy method and the method of complex moduli. Let us derive different relationships and compare them. This will give us an idea about the correlation of the results obtained by these two methods.

Bulk modulus of elasticity K is expressed in terms of Young's modulus, E, and shear modulus, G [19]:

$$K = \frac{EG}{3(3G - E)} \qquad (6.25)$$

To obtain the complex bulk modulus K^*, change E and G in Equation (6.25) to complex moduli $E^* = E' + iE''$ and $G^* = G' + iG''$ and then distinguish the real and the imaginary parts

$$K^* = K' + iK'' = \frac{3E'|G^*|^2 - G'|E^*|^2 + i(3E''|G^*|^2 - G''|E^*|^2)}{3[(3G' - E')^2 + (3G'' - E'')^2]}$$

$$(6.26)$$

Dissipation factor ψ'' is defined in terms of imaginary K'' and real K' parts of the complex bulk modulus

$$\psi'' = 2\pi K''/K'$$

or

$$\psi'' = 2\pi \frac{3E''|G^*|^2 - G''|E^*|^2}{3E'|G^*|^2 - G'|E^*|^2} \qquad (6.27)$$

It is possible to change imaginary parts of complex moduli to the relationships following from Equation (6.15):

$$E'' = \psi^+ E'/2\pi, \qquad G'' = \psi' G'/2\pi \qquad (6.28)$$

where ψ^+ and ψ' are the dissipation factors under uniaxial loading and ideal shear.

Applying Equations (6.27) and (6.28), one can derive the expression for dissipation factor ψ'' in the form

$$\psi'' = \frac{3\psi^+ G'\left(1 + \dfrac{\psi'^2}{4\pi^2}\right) - \psi' E'\left(1 + \dfrac{\psi^{+2}}{4\pi^2}\right)}{3G'\left(1 + \dfrac{\psi'^2}{4\pi^2}\right) - E'\left(1 + \dfrac{\psi^{+2}}{4\pi^2}\right)} \qquad (6.29)$$

Equation (6.29) relates the dissipation factor due to volume change to dissipation factors under uniaxial loading and ideal shear for the viscoelastic body.

Now let us consider energy losses, ΔW, under uniaxial loading on the basis of the energy method. The dissipation factor, ψ^+, and the amplitude of the elastic energy, W, determine energy losses

$$\Delta W = \psi^+ W \tag{6.30}$$

Total energy, W, can be divided into the volume change energy, W'', and the form change energy, W'. Energy losses are then expressed as:

$$\Delta W = \psi' W + \psi'' W'' \tag{6.31}$$

Comparison of Equations (6.30) and (6.31) under uniaxial tension (for example, along the axis x_1, $\sigma_{11} \neq 0$) makes it possible to derive the relationship between the coefficients ψ'', ψ^+, and ψ'.

It follows from Equations (5.21)–(5.26) that energy losses for the given type of loading are equal to

$$\Delta W = \psi^+ \sigma_{11}^2 / 2E$$

$$\Delta W = \frac{\psi' \sigma_{11}^2}{6G} + \frac{\psi'' \sigma_{11}^2}{18K} \tag{6.32}$$

Equating the right parts in Equations (6.32), one can obtain an expression for dissipation factor due to the volume change, ψ''

$$\psi'' = 3K \frac{3G\psi^+ - E\psi'}{EG}$$

or, considering Equation (5.27)

$$\psi'' = \frac{3G\psi^+ - E\psi'}{3G - E} \tag{6.33}$$

Now let us compare Equation (6.29), which corresponds to the method of complex moduli, and Equation (6.33), which corresponds to the energy method. Take into consideration the change from the real part of complex moduli E' and G' to the "elastic" moduli E and G. Also, ignore the terms that are on the order of the square of the dissipation factors, as they are of small value compared to the unity under poor energy dissipation. One can then see that Equation (6.29) coincides with Equation (6.33). The difference in Equations (6.29) and (6.33) follows from the change of the real part of the complex moduli to the elastic moduli on retention of imaginary parts. Such a change is not mathematically rigorous.

Let us compare the coefficients in Equations (6.23) and (6.24), using the relationships derived. The formula for coefficient A [Equation (5.43)] with consideration of Equation (6.33), will take the form:

$$A = \frac{\xi E_f \psi_f^+ + (1 - \xi)E_m \psi_m^+}{E_1^2} \qquad (6.34)$$

where ψ_f^+ and ψ_m^+ are the dissipation factors of the fiber and the matrix respectively under uniaxial loading.

The expression for the imaginary part of the complex compliance, s_{11}'', according to Equations (6.17) and (6.21) will then take the form:

$$s_{11}'' = -\frac{E_1''}{|E_1^*|^2} = -\frac{\xi E_f' \, \psi_f^+ + (1 - \xi)E_m' \psi_m^+}{2\pi E_1'^2[1 + (\xi E_f' \, \psi_f^+ + (1 - \xi)E_m' \psi_m^+)^2/4\pi^2 E_1'^2]} \qquad (6.35)$$

If one ignores the values in Equation (6.35), which are small as compared to the unity and changes the real parts of complex moduli to elastic moduli, then one can see that the coefficients from Equations (6.34) and (6.35) coincide to an accuracy of constant factors.

It is more convenient to compare composite dissipation factors, ψ_6^*, under ideal shear [Equations (5.37) and (6.22)], which were obtained by the two methods, than it is to compare the coefficients C and s_{66}'' from Equations (6.23) and (6.24). It follows from Equations (6.17) and (6.22) that

$$\psi_6^* = 2\pi = \frac{G_{12}''}{G_{12}'} = \frac{\xi G_m' \psi_f' \left(1 + \frac{\psi_m'^2}{4\pi^2}\right) + (1 - \xi)G_f' \, \psi_m' \left(1 + \frac{\psi_f'^2}{4\pi^2}\right)}{\xi G_m' \left(1 + \frac{\psi_m'^2}{4\pi^2}\right) + (1 - \xi)G_f' \left(1 + \frac{\psi_f'^2}{4\pi^2}\right)} \qquad (6.36)$$

The energy method results in the following formula for the dissipation factor, ψ_6^*:

$$\psi_6^* = \frac{\xi G_m \psi_6' + (1 - \xi)G_f \psi_m'}{\xi G_m + (1 - \xi)G_f} \qquad (6.37)$$

Compare Equations (6.36) and (6.37). The terms on the order of the square of the dissipation factors can be ignored, as they are of small value compared to the unity. One can then see that the energy method and the method of complex moduli result in analogous formulas in the case of ideal shear.

It can also be shown that the rest of the corresponding coefficients in the relationships for composite energy losses [Equations (6.23) and (6.24)] are approximately equal to each other. Additional estimation calculations were performed for different variants of the elasto-dissipative characteristics of composite constituents. It was shown that the corresponding coefficients in Equations (6.23) and (6.24) differ by no more than a few tenths of a percent.

Therefore, the results following from the two different methods of investigation of composite dissipative behavior (i.e., the method of complex moduli and the energy method) do not contradict each other. These methods can be used independently or they can supplement each other. In the end, expediency of application of one or another method is defined by the efficiency and simplicity of specific problems' solutions.

CHAPTER 7

Extreme Properties of the Dissipative Behavior of a Transversely Isotropic Monolayer (Unidirectional Composite)

The main parameter of energy dispersion in a material is the dissipation factor, ψ, which is the ratio of specific energy losses, ΔW, in a loading cycle to the amplitude value of the specific elastic energy, W. In cases when the energy losses, ΔW, are quadratic forms with respect to the stresses, the dissipation factor does not depend on the stress amplitude. At the same time, its values vary if the relation between normal and shear stresses varies, i.e., if the stress state type changes. Let us consider the effect of the stress state type on the relative energy dissipation of the unidirectional material.

7.1 SURFACE OF DISSIPATION

The dissipation factor of the transversely isotropic body under an arbitrary plane stress state is described, as it follows from Equations (4.55) and (4.93), by the formula

$$\psi(\sigma_1,\sigma_2,\tau_{12}) = \frac{\psi_{11}\sigma_1^2 + \psi_{22}\sigma_2^2 + 2\psi_{12}\sigma_1\sigma_2 + \psi_{66}\tau_{12}^2}{s_{11}\sigma_1^2 + s_{22}\sigma_2^2 + 2s_{12}\sigma_1\sigma_2 + s_{66}\tau_{12}^2} \qquad (7.1)$$

Assume that the stress amplitude $\sigma_1 \neq 0$, and then let us turn to non-dimensional stresses $\bar{\sigma}_2 = \sigma_2/\sigma_1$, $\bar{\tau}_{12} = \tau_{12}/\sigma_1$ in Equation (7.1). In this case

$$\psi(\bar{\sigma}_2,\bar{\tau}_{12}) = \frac{\psi_{11} + \psi_{22}\bar{\sigma}_2^2 + 2\psi_{12}\bar{\sigma}_2 + \psi_{66}\bar{\tau}_{12}^2}{s_{11} + s_{22}\bar{\sigma}_2^2 + 2s_{12}\bar{\sigma}_2 + s_{66}\bar{\tau}_{12}^2} \qquad (7.2)$$

131

Special cases of Equation (7.2) are the following:

- The stress $\bar{\sigma}_2 \rightarrow \infty$ and $\bar{\tau}_{12}$ is constrained, $\psi \rightarrow \psi_{22}/s_{22} = \psi_2^*$.
  Here ψ_2^* is the dissipation factor under uniaxial cyclic loading along the x_2-axis.
- The stress $\bar{\tau}_{12} \rightarrow \infty$ and $\bar{\sigma}_2$ is constrained, $\psi \rightarrow \psi_{66}/s_{66} = \psi_6^*$.

Here ψ_6^* is the dissipation factor under ideal shear in the axes (x_1, x_2).

Let us consider Equation (7.2) to be the equation of some surface in a three-dimensional space of Cartesian coordinates $(\bar{\tau}_{12}, \bar{\sigma}_2, \psi)$. Let us call the surface $\psi = \psi(\bar{\sigma}_2, \bar{\tau}_{12})$ the dissipation surface, and the $(\bar{\sigma}_2, \bar{\tau}_{12})$ plane the plane of nondimensional stresses. The dissipation surface is symmetrical with respect to the plane $\bar{\tau}_{12} = 0$ since the dissipation factor [Equation (7.2)] is the even function of $\bar{\tau}_{12}$. The dissipation surface for a real fibrous composite is shown in Figure 7.1. The horizontal coordinate plane is the plane of nondimensional stresses.

7.2 A CHART OF LEVEL LINES OF THE DISSIPATION SURFACE. EXTREME PROPERTIES OF RELATIVE ENERGY DISSIPATION

Let us analyze the geometry of the dissipation surface by examining the level lines, i.e., the sections of the surface with the planes $(\bar{\sigma}_2, \bar{\tau}_{12}, \psi = \psi^*)$ for different values of $\psi = \psi^* = $ const. The equations of level lines can be obtained using Equation (7.2), provided that $\psi = \psi^*$

$$(\psi_{22} - \psi^* s_{22})\bar{\sigma}_2^2 + (\psi_{12} - \psi^* s_{12})2\bar{\sigma}_2$$

$$+ (\psi_{66} - \psi^* s_{66})\bar{\tau}_{12}^2 + (\psi_{11} - \psi^* s_{11}) = 0 \qquad (7.3)$$

A set of level lines pictured on the same plane $(\bar{\sigma}_2, \bar{\tau}_{12})$ gives an idea of "the relief" of the dissipation surface. It can serve as a diagram that describes the dependence of the relative energy dissipation in the material upon the type of plane stress state. Let us call this diagram the chart of level lines of the dissipation surface.

The examples of the charts of level lines for real fibrous composites are illustrated by Figures 7.2 and 7.3. The figures on the level lines mean the value of the dissipation factor for the level line, $\psi^* = $ const. Dark rectangles show the direction of ψ decreasing. The charts of level lines were calculated with the help of special personal computer software.

If we exchange the components of the EDC strain matrix $[\psi]$ and the compliance matrix $[S]$ for their expressions in terms of elasto-dissipative

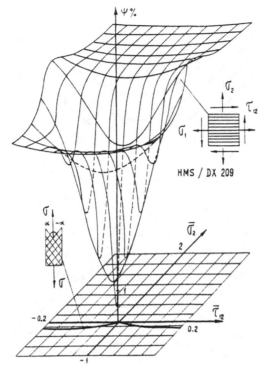

FIGURE 7.1. Dissipation surface of CFRP HMS/DX 209 in three-dimensional space.

constants [Equations (4.89) and (4.97)], Equation (7.3) will be of the form:

$$\frac{\psi_2^* - \psi^*}{E_2}\bar{\sigma}_2^2 + 2\left(\psi_{12} + \frac{\psi^*\nu_{12}}{E_1}\right)\bar{\sigma}_2$$

$$+ \frac{\psi_6^* - \psi^*}{G_{12}}\bar{\tau}_{12}^2 + \frac{\psi_1^* - \psi^*}{E_1} = 0 \qquad (7.4)$$

The component ψ_{12} retains its original form for the sake of simplicity, but its value can be calculated from Equations (4.90) and (4.91).

It might be well to point out the general properties of level lines of the dissipation surface [Equations (7.3) and (7.4)]:

(1) Level lines that conform to different values of the dissipation factor, ψ^*, cannot intersect as the function of Equation (7.2) is unambiguous.

(2) Level lines are curves of the second order.

Stationary points of the dissipation surface are at one time the special

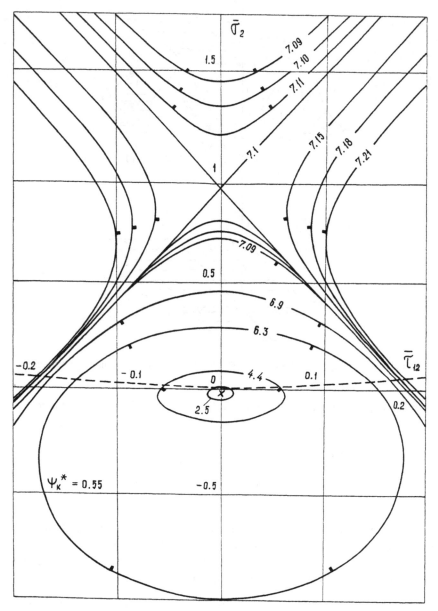

FIGURE 7.2. Chart of level lines of the dissipation surface shown in Figure 7.1 for CFRP HMS/DX 209.

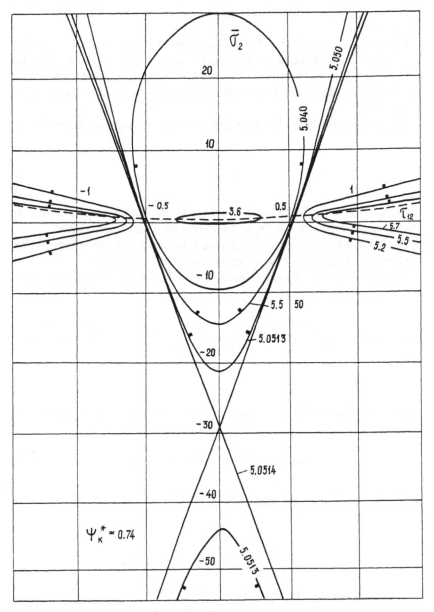

FIGURE 7.3. Chart of level lines of the dissipation surface for glass fiber reinforced plastic GLASS/DX 210.

points of a set of level lines where $\psi = $ const. A pair of intersecting straight lines, among the set of level lines [Equations (7.3) and (7.4)], has such a special point.

The particular type of this chart of level lines depends on the relation between the values of the elasto-dissipative engineering constants. For instance, let us investigate the sort of level lines that correspond to the three-component model of the unidirectional composite. In this case, $\psi_{12} = -\nu_{12}\psi_1^*/E_1$.

The following relation between engineering constants is believed to be true

$$\psi_1^* \leq \psi_2^* \leq \psi_6^* \tag{7.5}$$

where ψ_1^*, ψ_2^*, and ψ_6^* are the dissipation factors under loading along the fibers, transverse to the fibers, and under ideal shear respectively.

The behavior of real composites (see Chapter 3) usually meets the requirement of Equation (7.5), so let us consider only this case.

The equation of level lines [Equation (7.4)] is of the following form for the three-component model:

$$\frac{\psi_2^* - \psi^*}{E_2}\bar{\sigma}_2^2 - 2\frac{(\psi_1^* - \psi^*)\nu_{12}}{E_1}\bar{\sigma}_2 + \frac{\psi_6^* - \psi^*}{G_{12}}\bar{\tau}_{12}^2 + \frac{\psi_1^* - \psi^*}{E_1} = 0 \tag{7.6}$$

Equation (7.6) can be transformed to the canonical type of the second-order curve

$$\lambda_1(\bar{\sigma}_2 - \bar{\sigma}_2^0)^2 + \lambda_2\bar{\tau}_{12}^2 + g = 0 \tag{7.7}$$

where the coefficients are equal to

$$\lambda_1 = \frac{\psi_2^* - \psi^*}{E_2}, \qquad \lambda_2 = \frac{\psi_6^* - \psi^*}{G_{12}}$$

$$g = \frac{\psi_1^* - \psi^*}{E_1}\frac{\psi^*(\nu_{12}^2 E_2 - E_1) + E_1\psi_2^* - \psi_1^*\nu_{12}^2 E_2}{E_1\psi_2^* - E_1\psi^*} \tag{7.8}$$

$$\bar{\sigma}_2^0 = \frac{(\psi_1^* - \psi^*)\nu_{12}\sqrt{E_2}}{E_1\sqrt{\psi_2^* + \psi^*}}$$

Let $\psi^* < \psi_1^* \neq \psi_2^*$ (the case when $\psi_1^* = \psi_2^*$ will be examined singly). Then, in Equation (7.7), $\lambda_1 > 0, \lambda_2 > 0$. Take into account that the following inequality is true for elastic moduli and Poisson's ratios

$$E_1 > \nu_{12}^2 E_2$$

resulting from the condition of positive definiteness of the potential elastic energy. The condition for the free term g is then of the form: $g > 0$, which is why Equation (7.7) does not have a solution in real numbers.

Only one point fits the case when $\psi^* = \psi_1^*$; this is $\bar{\sigma}_2 = 0, \bar{\tau}_{12} = 0$.

The parameters of Equations (7.7) and (7.8) are as follows in the interval $\psi_1^* < \psi^* < \psi_2^* : \lambda_1 > 0, \lambda_2 > 0, g < 0$. Thus, level lines have the form of an ellipse (see Figures 7.4 and 7.5).

Let $\psi^* = \psi_2^*$. It follows that Equation (7.6) will be of the parabola form:

$$\lambda_1^0 \bar{\tau}_{12}^2 + 2h\bar{\sigma}_2 + f = 0 \qquad (7.9)$$

where

$$\lambda_1^0 = \frac{\psi_6^* - \psi_2^*}{G_{12}}, \qquad h = \frac{(\psi_2^* - \psi^*)\nu_{12}}{E_1}, \qquad f = \frac{\psi_1^* - \psi_2^*}{E_1}$$

Here, considering Equation (7.5) and $\psi_1^* \neq \psi_2^*$, the parameters $\lambda_1^0 > 0$, $h > 0, f < 0$. The $\bar{\sigma}_2$-axis is the symmetric axis for the parabola, Equation (7.9); its peak is located above the origin and its branches are directed to the negative side of the nondimensional stress axis, $\bar{\sigma}_2$ (see Figures 7.4 and 7.5). Level lines corresponding to different values of ψ^* cannot intersect. So all elliptical level lines fall inside the parabola in the interval $\psi_1^* < \psi^* < \psi_2^*$.

If $\psi^* > \psi_2^*$, Equation (7.6) of level lines is transformed to the form of Equation (7.4). The parameter, ψ_k^*, when $g = 0$, will play an important role in the further analysis. Let us call the parameter ψ_k^* the critical value of the dissipation factor.

Setting the numerator of the expression for g in Equation (7.8) equal to zero, one can obtain the expression for the critical value of the dissipation factor in terms of engineering elasto-dissipative constants

$$\psi_k^* = \frac{\psi_2^* E_1 - \psi_1^* \nu_{12}^2 E_2}{E_1 - \nu_{12}^2 E_2} \qquad (7.10)$$

The critical value of ψ_k^* when $\psi_1^* \neq \psi_2^*$ is always more than the value of the dissipation factor under loading transverse to the fiber direction, ψ_2^*, and is, at the same time, the stationary point of the dissipation surface, Equation (7.2).

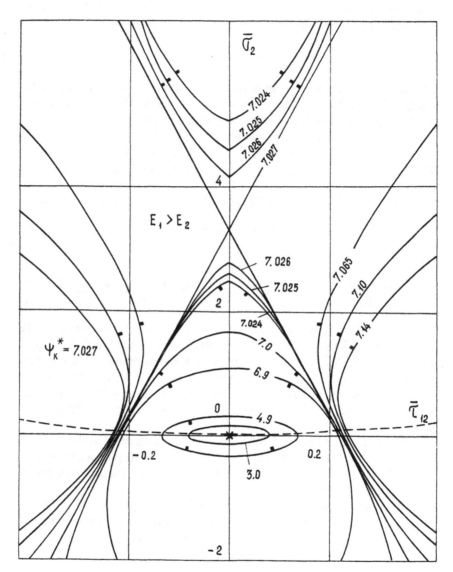

FIGURE 7.4. Chart of level lines of the dissipation surface for unidirectional fiber reinforced plastic in the case when $\psi_k^* < \psi_6^*$, $\psi_1^* \neq \psi_2^*$. Initial data for the three-component model of EDC are as follows: $E_1 = 200$ GPa, $E_2 = 10$ GPa, $G_{12} = 5$ GPa, $\nu_{12} = 0.3$, $\psi_1^* = 1\%$, $\psi_2^* = 7\%$, $\psi_6^* = 10\%$.

138

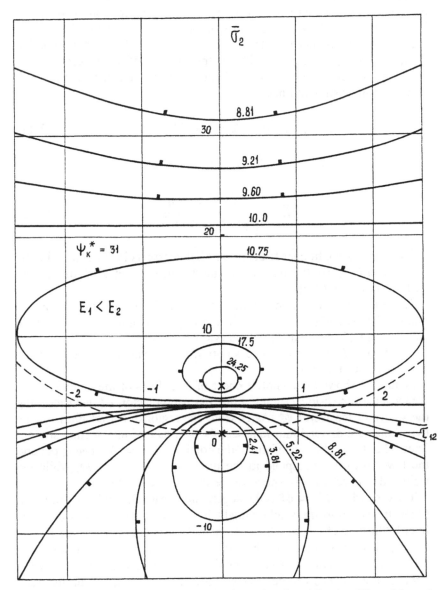

FIGURE 7.5. Chart of level lines of the dissipation surface for unidirectional fiber reinforced plastic in the case when $\psi_k^* > \psi_6^*$, $\psi_1^* \neq \psi_2^*$. Initial data for the three-component model of EDC are as follows: $E_1 = 10$ GPa, $E_2 = 200$ GPa, $G_{12} = 5$ GPa, $\nu_{12} = 0.2$, $\psi_1^* = 1\%$, $\psi_2^* = 7\%$, $\psi_6^* = 10\%$.

The subsequent analysis of level lines can proceed in two directions, depending on the relation between the value of the dissipation factor under ideal shear, ψ_δ^*, and the critical value of the dissipation factor, ψ_k^*.

Let $\psi_k^* < \psi_\delta^*$. In this case, for the interval $\psi_k^* < \psi^* < \psi_\delta^*$ in Equation (7.7), $\lambda_1 > 0$, $\lambda_2 < 0$, $g < 0$. This is the equation of the hyperbola.

At the particular value $\psi^* = \psi_k^*$, Equation (7.7) transforms to the equation of a pair of straight lines

$$\bar{\sigma}_2 = \pm \bar{K}_k \bar{\tau}_{12} + 1/\nu_{12} \tag{7.11}$$

Here the parameter $\bar{K}_k$ is calculated from the formula:

$$\bar{K}_k = \sqrt{\frac{(\psi_\delta^* - \psi_k^*)E_2}{[\psi_k^* - \psi_?^*]G_{12}}} \tag{7.12}$$

The pair of straight lines [Equation (7.11)] intersects in the point with the coordinates $(1/\nu_{12},0)$, this point being the stationary point of the dissipation surface [Equation (7.1)].

The equation of level lines, Equation (7.7), is the equation of the hyperbolas in the interval $\psi_k^* < \psi^* < \psi_\delta^*$, these hyperbolas are conjugate to the corresponding hyperbolas in the interval $\psi_?^* < \psi^* < \psi_k^*$.

Equation (7.7) does not have a solution for the values $\psi^* \geq \psi_\delta^*$, since $\lambda_1 \leq 0$, $\lambda_2 < 0$, $g < 0$.

Figure 7.4 shows the complete chart of level lines of the unidirectional composite for the critical values of the dissipation factor, ψ_k^*, not exceeding the values of the factor ψ_δ^*. As is seen, the regions of level lines for the hyperbolas are divided into four sectors by a pair of intersecting straight lines. The straight lines conform to the condition $\psi^* = \psi_k^*$. The upper and the lower sectors correspond to the interval $\psi_?^* < \psi^* < \psi_k^*$, while the right and left sectors correspond to the interval $\psi_k^* < \psi^* < \psi_\delta^*$.

The case when $\psi_k^* = \psi_\delta^*$ is the degeneration case. Level lines [Equation (7.11)] transform to the single straight line $\bar{\sigma}_2 = 1/\nu_{12}$, $\psi^* = \psi_\delta^*$ on the line. In the case of $\psi^* > \psi_\delta^*$, Equation (7.7) does not have a solution.

Let us investigate the case when the critical value of the dissipation factor, ψ_k^*, is more than the value of the dissipation factor, ψ_δ^*. The following interval of ψ^* values will be selected here: $\psi_?^* < \psi^* < \psi_\delta^*$. In this case, the parameters $\lambda_1 > 0$, $\lambda_2 < 0$, $g < 0$ in Equation (7.7), so level lines will be hyperbolas.

In the case of $\psi^* = \psi_\delta^*$, Equation (7.7) is transformed to the quadratic equation in $\bar{\sigma}_2$ with a positive discriminant and two different positive roots.

The following solutions are obtained:

$$\bar{\sigma}_2 = \cfrac{\cfrac{(\psi_6^* - \psi_1^*)E_2\nu_{12}}{(\psi_8^* - \psi_2^*)E_1} \pm \sqrt{\cfrac{\psi_6^* - \psi_1^*}{E_1^2 E_2}}[\cdots\cdots]}{\cfrac{\psi_6^* - \psi_2^*}{E_2}} \quad (7.13)$$

The level lines are the pair of parallel straight lines that lie in the positive half-plane above the $\bar{\tau}_{12}$-axis (see Figure 7.5).

In the interval $\psi_8^* < \psi^* < \psi_k^*$ the terms of Equation (7.7) are: $\lambda_1 < 0$, $\lambda_2 > 0$, $g > 0$. Here the ellipses are the level lines. As the value of ψ^* increases ($\psi^* \rightarrow \psi_k^*$), $\bar{\sigma}_2^0$ approaches $1/\nu_{12}$. The only isolated point with the coordinates $(1/\nu_{12},0)$ corresponds to the level line $\psi^* = \psi_k^*$.

In the case of $\psi^* > \psi_k^*$ a solution of Equation (7.7) does not exist.

Construction of the chart of level lines of the dissipation surface was performed for different values of the dissipation factors $\psi_1^* \neq \psi_2^*$. If these coefficients coincide, but the Young's moduli differ ($E_1 \neq E_2$), then the critical value of the dissipation factor equals

$$\psi_k^* = \psi_1^* \quad (7.14)$$

Analysis of the equation of the level lines [Equation (7.7)] shows that there is no solution when $\psi^* < \psi_1^*$ and $\psi^* \geq \psi_8^*$. If $\psi^* = \psi_1$, then the level line is the straight line

$$\bar{\tau}_{12} = 0$$

In the range $\psi_1^* < \psi^* < \psi_8^*$ hyperbolas are the level lines, which are symmetrical with respect to the $\bar{\sigma}_2$-axis [$\lambda_1 > 0$, $\lambda_2 < 0$, $g > 0$ in Equation (7.7)]. Figure 7.6 gives the chart of the level lines of the dissipation surface for the case under consideration.

Let us draw the chart of the level lines of the dissipation surface for the isotropic material in the case where two constants define its dissipative behavior. Equation (7.2) of the dissipation surface takes the form

$$\psi = \frac{\psi_1^*\bar{\sigma}_2^2 + 2[\psi_1^* - \psi_8^*(1 + \nu)]\bar{\sigma}_2 + 2(1 + \nu)\psi_8^*\bar{\tau}_{12}^2 + \psi_1^*}{\bar{\sigma}_2^2 - 2\nu\bar{\sigma}_2 + 2(1 + \nu)\bar{\tau}_{12}^2 + 1}$$

$$(7.15)$$

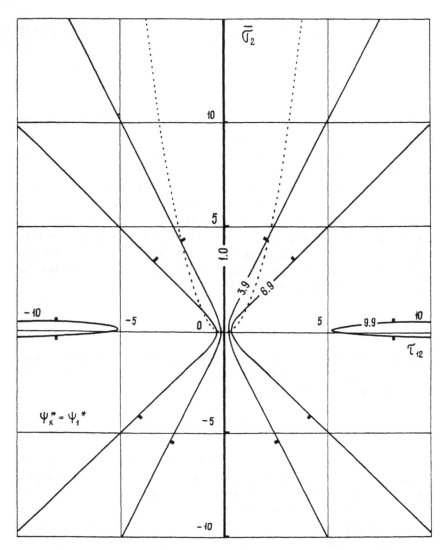

FIGURE 7.6. Chart of level lines of the dissipation surface for unidirectional fiber reinforced plastic in the case when $\psi_1^* = \psi_2^*$. Initial data for the three-component model of EDC are as follows: $E_1 = 200$ GPa, $E_2 = 10$ GPa, $G_{12} = 5$ GPa, $\nu_{12} = 0.3$, $\psi_1^* = \psi_2^* = 1\%$, $\psi_6^* = 10\%$.

The equation that corresponds to Equation (7.15) for the level lines, $\psi = \psi^*$, is of the form

$$(\psi_1^* - \psi^*)\bar{\sigma}_2^2 + 2[\psi_1^* + \nu\psi^* - \psi_\delta^*(1 + \nu)]\bar{\sigma}_2$$

$$+ 2(1 + \nu)(\psi_\delta^* - \psi^*)\bar{\tau}_{12}^2 + (\psi_1^* - \psi^*) \qquad (7.16)$$

As in the case with the fibrous composite, let us straighten out the dissipation factor values, taking $\psi_1^* < \psi_\delta^*$. Experimental data [93] provide support for such a relationship.

Let us study different variants of the ψ^* values. If $\psi^* = \psi_1^*$, the parabola is a level line which does not depend upon EDC, i.e.,

$$\bar{\sigma}_2 = \bar{\tau}_{12}^2 \qquad (7.17)$$

Let us transform Equation (7.16) for the case $\psi^* \neq \psi_1^*$. Taking the perfect square with respect to $\bar{\sigma}_2$ in the form of Equation (7.7), one then obtains

$$\lambda_1\bar{\tau}_{12}^2 + \lambda_2(\bar{\sigma}_2 - \bar{\sigma}_2^0)^2 + g = 0 \qquad (7.18)$$

where

$$\lambda_1 = 2(1 + \nu)(\psi_\delta^* - \psi^*)$$

$$\lambda_2 = \psi_1^* - \psi^*, \qquad \bar{\sigma}_2^0 = \frac{\psi_1^* + \nu\psi^* - \psi_\delta^*(1 + \nu)}{\psi^* - \psi_1^*}$$

$$g = (\psi_1^* - \psi^*) - \frac{[\psi_1^* + \nu\psi^* - \psi_\delta^*(1 + \nu)]^2}{\psi_1^* - \psi^*}$$

or

$$g = \frac{(1 + \nu)(\psi_\delta^* - \psi^*)[2\psi_1^* + \psi^*(\nu - 1) - \psi_\delta^*(1 + \nu)]}{\psi_1^* - \psi^*}$$

If $\psi^* < \psi_1^*$, the terms in Equation (7.18) are: $\lambda_1 > 0$, $\lambda_2 > 0$ and the sign of the term g is defined by the position of the point, ψ_k^*. The parameter, ψ_k^*, is the dissipation factor when $g = 0$. The value of ψ_k^* is calculated from the following formula:

$$\psi_k^* = \frac{2\psi_1^* - \psi_\delta^*(1 + \nu)}{1 - \nu} \qquad (7.19)$$

A nonnegative value of ψ_k^* always exists, as the numerator in Equation (7.19) is always nonnegative (the condition of nonnegative definiteness of the square form of the matrix $[\psi]$), and $\nu \leq 0.5$. If $\psi^* < \psi_x^*$, then it is always the case that $\psi_k^* < \psi_1^*$. The parameter ψ_k^* defined from Equation (7.19) is known as the critical dissipation factor of the isotropic body. If $\psi^* < \psi_k^*$, then $g > 0$ in Equation (7.18) and the equation does not have a solution. This means that ψ_k^* is the lower bound for variation of the dissipation factor of the isotropic body.

If $\psi^* = \psi_k^*$, then the only point with the coordinates ($\bar{\tau}_{12} = 0$, $\bar{\sigma}_2 = 1$) is the solution of Equation (7.18). Note that if there is a relation between the dissipation factors of the isotropic body in the form $\psi_\delta^* = 2\psi_1^*/(1 + \nu)$, the body does not disperse energy under uniform cyclic biaxial loading ($\psi = 0$), although the dissipation factors under uniaxial loading and shear differ from zero.

The level lines are ellipses in the interval $\psi_k^* < \psi^* < \psi_1^*$. Equation (7.18) has the following parameters in the interval $\psi_1^* < \psi^* < \psi_\delta^*$:$\lambda_1 > 0$, $\lambda_2 < 0$, $g > 0$. For this reason, the level lines are hyperbolas in this interval.

In the case when $\psi^* = \psi_\delta^*$, the level line is a straight line of the form: $\bar{\sigma}_2 = -1$ [the solution of Equation (7.18)]. But if $\psi^* > \psi_\delta^*$, then the parameters $\lambda_1 < 0$, $\lambda_2 < 0$, $g < 0$ in Equation (7.18), i.e., there is no solution in this case. The chart of level lines for the isotropic body is shown in Figure 7.7.

Analysis of the chart [Figure (7.7)] allows one to conclude that the values of the dissipation factor fall within the interval $[\psi_k^*, \psi_\delta^*]$ for the isotropic body under an arbitrary plane stress state, where ψ_k^* is the critical value of the dissipation factor [Equation (7.19)]. The critical value of the dissipation factor corresponds to the uniform biaxial cyclic loading of the body and always meets the requirement: $\psi_1^* > \psi_k^*$.

7.3 TRAJECTORIES OF STRESS STATE VARIATION

A change in the material stress state can be reflected on the plane of nondimensional stress amplitudes ($\bar{\sigma}_2, \bar{\tau}_{12}$) by a curved line that will be called the trajectory of stress state variation (TSSV). The chart of level lines of the dissipation surface with the TSSV drawn in it gives an indication of the function type. The point of intersection of the level line and the TSSV conforms to the specific value of the material dissipation factor.

Write the rules for stress state variation in the coordinate system related to the body when rotating the system through an angle, α, with respect to the coordinate axes of the fixed external stress state. Then, using Equation

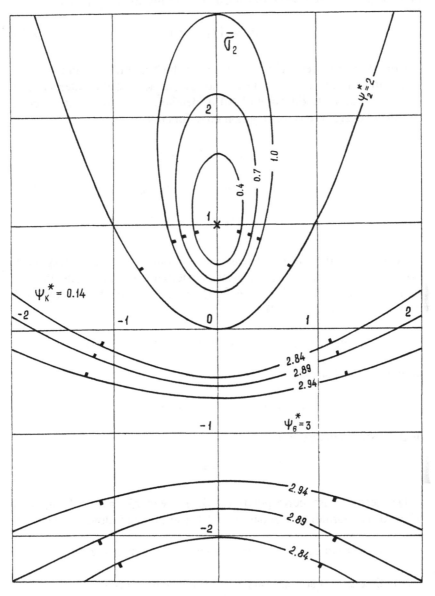

FIGURE 7.7. Chart of level lines of the dissipation surface for the isotropic body under arbitrary plane stress state. Initial data are as follows: E = 200 GPa, ν = 0.3, ψ_i^* = 2%, ψ_s^* = 3%.

(4.105), perform an inverse transformation of the stress matrices:

$$\{\sigma\} = [T_1]^{-1}\{\sigma'\} \tag{7.20}$$

where $\{\sigma\}$ are the stress column matrices in the "natural" coordinate system of the monolayer, and $\{\sigma'\}$ are the stress matrices in the external fixed system. Next, write Equation (7.20) in detailed form, following accepted shorthand designations for trigonometric functions

$$\sigma_1 = \sigma_1' \, c^2 + \sigma_2' \, s^2 + \tau_{12}' 2 \, sc$$

$$\sigma_2 = \sigma_1' \, s^2 + \sigma_2' \, c^2 - \tau_{12}' 2 \, sc \tag{7.21}$$

$$\tau_{12} = (\sigma_2' - \sigma_1') \, sc + \tau_{12}' \, (c^2 - s^2)$$

Consider the trajectories of stress state variation, which correspond to Equation (7.21) under particular cases of external fixed stress state (the constant vector $\{\sigma'\}$). Let $\sigma_1' = \sigma \neq 0$, $\sigma_2' = 0$, $\tau_{12}' = 0$ (the uniaxial loading), then in Equation (7.21)

$$\sigma_1 = c^2 \, \sigma, \qquad \sigma_2 = s^2 \, \sigma, \qquad \tau_{12} = -sc \, \sigma$$

nondimensional stresses

$$\bar{\sigma}_2 = tg^2 \, \alpha, \qquad \bar{\tau}_{12} = -tg \, \alpha$$

or the TSSV is the parabola

$$\bar{\sigma}_2 = \bar{\tau}_{12}^2 \tag{7.22}$$

The second case corresponds to ideal shear $\tau_{12}' = \tau \neq 0$, $\sigma_1' = 0$, $\sigma_2' = 0$. The formulas of Equation (7.21) result in the following equations:

$$\sigma_1 = \tau \sin 2\alpha, \qquad \sigma_2 = -\tau \sin 2\alpha, \qquad \tau_{12} = \tau \cos 2\alpha$$

Nondimensional stresses vary in accordance to the rules

$$\bar{\sigma}_2 = -1, \qquad \bar{\tau}_{12} = ctg \, 2\alpha, \qquad -\infty < ctg \, 2\alpha < \infty$$

This means that the TSSV is a straight line parallel to the $\bar{\tau}_{12}$-axis, i.e.,

$$\bar{\sigma}_2 = -1 \tag{7.23}$$

For an isotropic body the TSSVs which conform to Equation (7.21) [or to Equations (7.22) and (7.23) in particular cases] coincide with the level lines of the dissipation surface.

Figure 7.4–7.6 illustrate the charts of level lines of the dissipation surface for unidirectional composites. The TSSVs [Equation (7.22)] are shown by dashed lines. As can be seen, the TSSV [Equations (7.22) and (7.23)] intersects the level lines. The position of the intersection points determines the rule of dissipation factor varying under uniaxial loading and ideal shear. Moving along the TSSV, one can see that the parabola [Equation (7.22)] passes through the level lines region with increased value of dissipation factors, while the straight line [Equation (7.23)] intersects the level line with the lowest dissipation factor value on the $\bar{\sigma}_2$-axis. Therefore dissipation factors reach a maximum in the case of uniaxial loading, while they reach a minimum value under ideal shear in the point conforming to the rotation angle, $\pi/4$.

CHAPTER 8

Dissipative Properties of Multilayered Composites

It is customary to assume that multilayered composite materials are materials made by a successive stacking of several different monolayers (also known as plies, layers, or laminae) (see Figure 8.1). In the general case each monolayer is orthotropic, distinguished by its own set of characteristics: $E_1^{(k)}$, $E_2^{(k)}$, $G_{12}^{(k)}$, $\nu_{12}^{(k)}$, and $h^{(k)}$. Here k is the layer number in the n-layered laminate, while $h^{(k)}$ is the layer thickness. We will be concerned with the following coordinate systems (see Figure 8.2): the general or global coordinate system (x_1, x_2) for the multilayered composite, and the local or natural coordinate systems for the monolayers $(x_1^{(k)}, x_2^{(k)})$. The axes of the local coordinates coincide with the monolayers' main axes of orthotropy. If the kth monolayer is the unidirectional composite, then the $x_1^{(k)}$-axis is directed along the reinforcing fibers in the material. The angle $\alpha^{(k)}$ indicates the orientation of the main axes of the kth monolayer in the global coordinate system.

8.1 ELASTIC CHARACTERISTICS OF MULTILAYERED COMPOSITES

This chapter will use the following nomenclature:

- stresses and strains in the kth monolayer in the direction of the main (natural) monolayer axes $(x_1, x_2)^{(k)}$

$$\{\sigma\}^{(k)} = \begin{Bmatrix} \sigma_1^0 \\ \sigma_2^0 \\ \tau_{12}^0 \end{Bmatrix}^{(k)}, \qquad \{\epsilon\}^{(k)} = \begin{Bmatrix} \epsilon_1^0 \\ \epsilon_2^0 \\ \gamma_{12}^0 \end{Bmatrix}^{(k)}$$

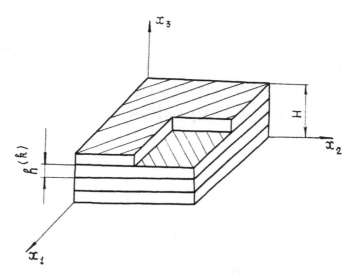

FIGURE 8.1. A multilayered composite material.

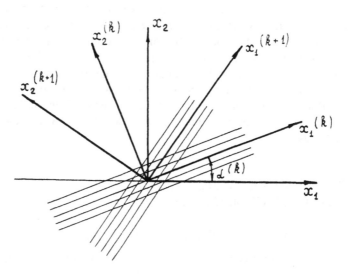

FIGURE 8.2. Monolayer natural coordinate system $(x_1^{(k)}, x_2^{(k)})$ and laminate global coordinate system (x_1, x_2).

- stresses and strains in the kth monolayer in the direction of the general (global) laminate coordinate axes (x_1, x_2)

$$\{\sigma\}^{(k)} = \begin{Bmatrix} \sigma_1 \\ \sigma_2 \\ \tau_{12} \end{Bmatrix}^{(k)}, \qquad \{\epsilon\}^{(k)} = \begin{Bmatrix} \epsilon_1 \\ \epsilon_2 \\ \gamma_{12} \end{Bmatrix}^{(k)}$$

- mean laminate stresses and strains in the direction of the general (global) laminate coordinate axes (x_1, x_2)

$$\{\sigma\} = \begin{Bmatrix} \sigma_1 \\ \sigma_2 \\ \tau_{12} \end{Bmatrix}, \qquad \{\epsilon\} = \begin{Bmatrix} \epsilon_1 \\ \epsilon_2 \\ \gamma_{12} \end{Bmatrix}$$

8.1.1 Plane Stress State of the Multilayered Composite

Let us assume that the layers deform integrally without slipping, i.e., corresponding strains of all layers are the same and are equal to the mean strains of the multilayered composite

$$\epsilon_1 = \epsilon_1^{(k)}, \quad \epsilon_2 = \epsilon_2^{(k)}, \quad \gamma_{12} = \gamma_{12}^{(k)} \qquad (k = 1,2,\ldots,n) \qquad (8.1)$$

Mean stresses of the laminate are defined in the following way:

$$\{\sigma\} = \begin{Bmatrix} \sigma_1 \\ \sigma_2 \\ \tau_{12} \end{Bmatrix} = \begin{Bmatrix} \displaystyle\sum_{k=1}^{n} \sigma_1^{(k)} \bar{h}^{(k)} \\[2mm] \displaystyle\sum_{k=1}^{n} \sigma_2^{(k)} \bar{h}^{(k)} \\[2mm] \displaystyle\sum_{k=1}^{n} \tau_{12}^{(k)} \bar{h}^{(k)} \end{Bmatrix} \qquad (8.2)$$

Here $\bar{h}^{(k)}$ is the relative thickness of the kth layer

$$\bar{h}^{(k)} = h^{(k)}/H$$

where H is the total laminate thickness

$$H = \sum_{k=1}^{n} \bar{h}^{(k)}$$

Substituting Hooke's law for the monolayer in Equation (4.28), one obtains

$$\{\sigma\} = [G]\{\epsilon\} \tag{8.3}$$

or

$$\begin{Bmatrix} \sigma_1 \\ \sigma_2 \\ \tau_{12} \end{Bmatrix} = \begin{bmatrix} g_{11} & g_{12} & g_{16} \\ g_{12} & g_{22} & g_{26} \\ g_{16} & g_{26} & g_{66} \end{bmatrix} \begin{Bmatrix} \epsilon_1 \\ \epsilon_2 \\ \tau_{12} \end{Bmatrix}$$

where

$$g_{ij} = \sum_{k=1}^{n} g_{ij}^{(k)} \bar{h}^{(k)} \qquad (i,j = 1,2,6) \tag{8.4}$$

Inverting the relationship (8.3)

$$\{\epsilon\} = [S]\{\sigma\} \tag{8.5}$$

Here $[G]$ and $[S]$ are the stiffness and the compliance matrices of the laminate

$$[S] = [G]^{-1} \tag{8.6}$$

The coefficients of the compliance matrix $[S]$ relate to the coefficients of the stiffness matrix $[G]$, which are defined by Equation (8.4), as follows:

$$S_{11} = \frac{g_{22}g_{66} - g_{26}^2}{\Delta g} \; ; \qquad S_{12} = \frac{g_{16}g_{26} - g_{66}g_{12}}{\Delta g}$$

$$S_{22} = \frac{g_{11}g_{66} - g_{16}^2}{\Delta g} \; ; \qquad S_{16} = \frac{g_{12}g_{26} - g_{22}g_{16}}{\Delta g} \tag{8.7}$$

$$S_{66} = \frac{g_{11}g_{22} - g_{12}^2}{\Delta g} \; ; \qquad S_{26} = \frac{g_{12}g_{16} - g_{11}g_{26}}{\Delta g}$$

$$\Delta g = \det [G]$$

Engineering constants of the multilayered composite—moduli of elasticity in the directions of the x_1- and x_2-axes, E_1 and E_2, shear modulus, G_{12}, and Poisson's ratio, ν_{12}—are expressed in the global coordinate system in the following way:

$$E_1 = \frac{\Delta g}{g_{22}g_{66} - g_{26}^2}\,; \qquad E_2 = \frac{\Delta g}{g_{11}g_{66} - g_{16}^2}$$

$$G_{12} = \frac{\Delta g}{g_{11}g_{22} - g_{12}^2}\,; \qquad \nu_{12} = \frac{g_{12}g_{66} - g_{16}g_{26}}{g_{22}g_{66} - g_{26}^2}$$

(8.8)

In the case of the orthotropic laminate ($g_{16} = g_{26} = 0$) Equations (8.8) becomes simpler

$$E_1 = g_{11} - g_{12}^2/g_{22}; \qquad E_2 = g_{22} - g_{12}^2/g_{11}$$

$$G_{12} = g_{66}; \qquad \nu_{12} = g_{12}/g_{22}$$

(8.9)

8.1.2 Bending of Multilayered Composites

Let us assume that the layers are bonded ideally (i.e., there is no mutual slipping of the layers in the laminate). Classical plate theory based on Kirchhoff-Love hypotheses (see, for example, References [8] and [45]) results in the following formulas for laminate strains:

$$\{\epsilon\} = [G]\{e\} + z\{\varkappa\}$$

(8.10)

or

$$\epsilon_1 = e_1 + z\varkappa_1$$

$$\epsilon_2 = e_2 + z\varkappa_2$$

$$\gamma_{12} = e_{12} + z\varkappa_{12}$$

Here z is the distance along the x_3 direction from a coordinate plane. Any plane that is parallel to the bounds of the laminate layers may be considered as such a coordinate plane (see Figure 8.3), $\{e\}$ is the strain vector of the

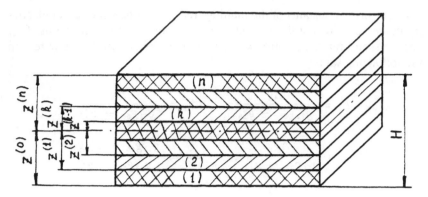

FIGURE 8.3. An n-layered laminate.

coordinate plane, and $\{x\}$ is the vector of laminate curvature variation. These matrices are of the forms:

$$\{\epsilon\} = \begin{Bmatrix} \epsilon_1 \\ \epsilon_2 \\ \gamma_{12} \end{Bmatrix}, \qquad \{e\} = \begin{Bmatrix} e_1 \\ e_2 \\ e_{12} \end{Bmatrix}, \qquad \{x\} = \begin{Bmatrix} x_1 \\ x_2 \\ x_{12} \end{Bmatrix} \qquad (8.11)$$

Let us introduce the matrix of the forces $\{T\}$ and the matrix of the moments $\{M\}$, which are related to the unit length of the coordinate line. These column matrices have the components:

$$\{T\} = \begin{Bmatrix} T_1 \\ T_2 \\ T_{12} \end{Bmatrix}, \qquad \{M\} = \begin{Bmatrix} M_1 \\ M_2 \\ M_{12} \end{Bmatrix} \qquad (8.12)$$

where the values of the components are calculated in terms of stresses in the monolayers as:

$$\{T\} = \sum_{k=1}^{n} \int_{z^{(k-1)}}^{z^{(k)}} \{\sigma\}^{(k)} dz$$

$$\{M\} = \sum_{k=1}^{n} \int_{z^{(k-1)}}^{z^{(k)}} \{\sigma\}^{(k)} z \, dz \qquad (8.13)$$

Use of integral sums in Equation (8.13) makes it possible to consider a partially linear character of stresses distribution along the laminate thickness. The expression for stresses in the layer [Equation (8.3)] with consideration of Equation (8.10) will be written as:

$$\{\sigma\}^{(k)} = [G]^{(k)}(\{e\} + z\{x\}) \tag{8.14}$$

Substituting Equation (8.14) into Equation (8.13) gives

$$\{T\} = \sum_{k=1}^{n} \int_{z^{(k-1)}}^{z^{(k)}} [G]^{(k)} dz\{e\} + \sum_{k=1}^{n} \int_{z^{(k-1)}}^{z^{(k)}} [G]^{(k)} z\, dz\{x\} \tag{8.15}$$

$$\{M\} = \sum_{k=1}^{n} \int_{z^{(k-1)}}^{z^{(k)}} [G]^{(k)} z\, dz\{e\} + \sum_{k=1}^{n} \int_{z^{(k-1)}}^{z^{(k)}} [G]^{(k)} z^2 dz\{x\}$$

or

$$\begin{Bmatrix} T \\ M \end{Bmatrix} = \begin{bmatrix} [G_1] & [G_2] \\ [G_2] & [G_3] \end{bmatrix} \begin{Bmatrix} e \\ x \end{Bmatrix} \tag{8.16}$$

Here the submatrices with 3×3 dimension $[G_j]$ ($j = 1,2,3$) are expressed in terms of monolayer stiffness matrices with the formulas

$$[G_j] = \sum_{k=1}^{n} \int_{z^{(k-1)}}^{z^{(k)}} [G]^{(k)} z^{j-1} dz \qquad (j = 1,2,3) \tag{8.17}$$

The following designations are often used in the literature to describe the matrices $[G_j]$:

$$[G_1] = [A], \qquad [G_2] = [C], \qquad [G_3] = [D] \tag{8.18}$$

Then Equation (8.16) takes the form

$$\begin{Bmatrix} T \\ M \end{Bmatrix} = \begin{bmatrix} [A] & [C] \\ [C] & [D] \end{bmatrix} \begin{Bmatrix} e \\ x \end{Bmatrix} \tag{8.16a}$$

or in the detailed form

$$\begin{Bmatrix} T_1 \\ T_2 \\ T_{12} \\ M_1 \\ M_2 \\ M_{12} \end{Bmatrix} = \begin{bmatrix} A_{11} & A_{12} & A_{16} & C_{11} & C_{12} & C_{16} \\ A_{12} & A_{22} & A_{26} & C_{12} & C_{22} & C_{26} \\ A_{16} & A_{26} & A_{66} & C_{16} & C_{26} & C_{66} \\ C_{11} & C_{12} & C_{16} & D_{11} & D_{12} & D_{16} \\ C_{12} & C_{22} & C_{26} & D_{12} & D_{22} & D_{26} \\ C_{16} & C_{26} & C_{66} & D_{16} & D_{26} & D_{66} \end{bmatrix} \begin{Bmatrix} e_1 \\ e_2 \\ e_{12} \\ x_1 \\ x_2 \\ x_{12} \end{Bmatrix} \tag{8.19}$$

If the components of the matrix $[G]^{(k)}$ do not vary with the thickness of the kth layer, then performing the integration in Equation (8.17), one obtains

$$[G_j] = \frac{1}{j} [G]^{(k)}(z^j_{(k)} - z^j_{(k-1)}) \qquad (8.20)$$

where $j = 1,2,3$; $k = 1,2,\ldots,n$.

The specific forms of the matrices $[A]$, $[C]$, and $[D]$ ($[G_j]$) depend on the laminate structure and the position of the coordinate plane.

8.1.2.1 EXAMPLE 1. A SYMMETRICAL LAMINATE STRUCTURE WITH AN ODD TOTAL NUMBER OF LAYERS

The scheme of this type of material is shown in Figure 8.4(a). All of the layers are equal in thickness and properties. A section of monolayers [two

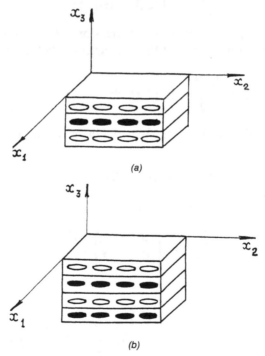

(a)

(b)

FIGURE 8.4. Examples of multilayered composites: (a) symmetrical laminate structure with odd total number of the layers, and (b) asymmetrical laminate structure with even total number of the layers.

monolayers in Figure 8.4(a)] is arranged so that their fibers form the angle $+\alpha$ to the direction of the global laminate axis, x_1, while the other section of monolayers forms the angle, $-\alpha$. Light fiber sections in the figure conform to the layers with the orientation angle, $+\alpha$, dark fiber sections conform to the layers with the orientation angle, $-\alpha$. In this case the coefficients of the stiffness submatrices $[A]$, $[C]$, and $[D]$ are expressed as

$$(A_{11}, A_{12}, A_{22}, A_{66}) = (g_{11}^{(k)}, g_{12}^{(k)}, g_{22}^{(k)}, g_{66}^{(k)})H$$

$$(A_{16}, A_{26}) = (g_{16}^{(k)}, g_{26}^{(k)})H/n$$

$$C_{ij} = 0 \qquad (i,j = 1,2,6)$$

$$(D_{11}, D_{12}, D_{22}, D_{66}) = (g_{11}^{(k)}, g_{12}^{(k)}, g_{22}^{(k)}, g_{66}^{(k)})H^3/12$$

$$(D_{16}, D_{26}) = (g_{16}^{(k)}, g_{26}^{(k)})\frac{3n^2 - 2}{n^3}\frac{H^3}{12}$$

The components A_{16}, A_{26}, D_{16}, and D_{26} have the sign of the orientation angle of the outer laminate layers. The other nonzero components are positive.

8.1.2.2 EXAMPLE 2. AN ASYMMETRICAL LAMINATE STRUCTURE WITH AN EVEN TOTAL NUMBER OF LAYERS

The scheme of the material is illustrated by Figure 8.4(b). In this case the coefficients of the stiffness submatrices $[A]$, $[C]$, and $[D]$ in Equation (8.19) are of the form

$$(A_{11}, A_{12}, A_{22}, A_{66}) = (g_{11}^{(k)}, g_{12}^{(k)}, g_{22}^{(k)}, g_{66}^{(k)})H$$

$$A_{16} = A_{26} = 0$$

$$C_{11} = C_{12} = C_{22} = C_{66} = 0$$

$$(C_{16}, C_{26}) = (g_{16}^{(k)}, g_{26}^{(k)})H^2/2n$$

$$(D_{11}, D_{12}, D_{22}, D_{66}) = (g_{11}^{(k)}, g_{12}^{(k)}, g_{22}^{(k)}, g_{66}^{(k)})H^3/12$$

$$(D_{16}, D_{26}) = 0$$

The components, C_{16}, C_{26}, have the sign of the orientation angle of the outer laminate layers, if z is positive. The other nonzero components are positive.

8.1.3 Inverse Relations

The relationships (8.19) can be inverted. If partial inversion takes place, then

$$\begin{Bmatrix} e \\ M \end{Bmatrix} = \begin{bmatrix} A' & C_1' \\ C_2' & D' \end{bmatrix} \begin{Bmatrix} T \\ x \end{Bmatrix} \tag{8.21}$$

or

$$\begin{Bmatrix} T \\ x \end{Bmatrix} = \begin{bmatrix} A'' & C_1'' \\ C_2'' & D'' \end{bmatrix} \begin{Bmatrix} e \\ M \end{Bmatrix} \tag{8.22}$$

where

$$[A'] = [A]^{-1};$$

$$[C_1'] = -[A]^{-1}[C]$$

$$[C_2'] = [C][A]^{-1};$$

$$[D'] = [D] - [C][A]^{-1}[C]$$

$$[A''] = [A] - [C][D]^{-1}[C];$$

$$[C_1''] = [C][D]^{-1}$$

$$[C_2''] = -[D]^{-1}[C];$$

$$[D''] = [D]^{-1}$$

The total inversion of Equation (8.19) gives

$$\begin{Bmatrix} e \\ x \end{Bmatrix} = \begin{bmatrix} A* & C* \\ C* & D* \end{bmatrix} \begin{Bmatrix} T \\ M \end{Bmatrix} \tag{8.23}$$

or

$$\begin{Bmatrix} e \\ x \end{Bmatrix} = \begin{bmatrix} [S_1] & [S_2] \\ [S_2] & [S_3] \end{bmatrix} \begin{Bmatrix} T \\ M \end{Bmatrix}$$

where

$$[A*] = [K]^{-1}; \qquad [C*] = -[K]^{-1}[C][D]^{-1}$$

$$[D*] = -[D]^{-1} + [D]^{-1}[C][K]^{-1}[C][D]^{-1}$$

$$[K] = [A] - [C][D]^{-1}[C]$$

If the material has a symmetrical structure and the mid-plane is selected as the coordinate plane, then Equations (8.23) become much simpler:

$$[A*] = [A]^{-1}; \qquad [C*] = [0]; \qquad [D*] = [D]^{-1}$$

The potential energy of deformation of the multilayered plates under bending is described by the formulas

$$W = \frac{1}{2} \begin{Bmatrix} T \\ M \end{Bmatrix}^{\mathrm{T}} \begin{bmatrix} [S_1] & [S_2] \\ [S_2] & [S_3] \end{bmatrix} \begin{Bmatrix} T \\ M \end{Bmatrix}$$

or

$$W = \frac{1}{2} \begin{Bmatrix} e \\ \varkappa \end{Bmatrix}^{\mathrm{T}} \begin{bmatrix} [G_1] & [G_2] \\ [G_2] & [G_3] \end{bmatrix} \begin{Bmatrix} e \\ \varkappa \end{Bmatrix} \qquad (8.24)$$

or

$$W = \frac{1}{2} \begin{Bmatrix} e \\ \varkappa \end{Bmatrix}^{\mathrm{T}} \begin{Bmatrix} T \\ M \end{Bmatrix}$$

For symmetrical structures

$$[S_2] = [0], \qquad [G_2] = [0]$$

8.2 DISSIPATIVE BEHAVIOR OF MULTILAYERED COMPOSITES UNDER A PLANE STRESS STATE

Energy loss in multilayered composite with ideal layers bonding under cyclic loading is equal to the sum of the energy losses in the layers. The values of the losses during the loading cycle in the kth layer of the composite are determined (1) by the EDC strain matrix $[\varphi]^{(k)}$ in terms of the amplitude values of strain vector components, or (2) with the help of the EDC stress matrix $[\psi]^{(k)}$ in terms of the amplitude values of the stresses, or (3) with the help of the mixed matrix $[\chi]^{(k)}$ [Equation (4.93)]. It is therefore possible to represent energy losses in a multilayered parallelepiped of unit length and width related to height H in three forms:

$$\Delta W = \frac{1}{2} \sum_{i=1}^{n} \bar{h}^{(i)} \{\epsilon\}^{\mathrm{T}} [\varphi^{(i)}] \{\epsilon\}$$

$$\Delta W = \frac{1}{2} \sum_{i=1}^{n} \bar{h}^{(i)} \{\epsilon\}^{\mathrm{T}} [\chi^{(i)}] \{\sigma^{(i)}\} \qquad (8.25)$$

$$\Delta W = \frac{1}{2} \sum_{i=1}^{n} \bar{h}^{(i)} \{\sigma^{(i)}\}^{\mathrm{T}} [\psi^{(i)}] \{\sigma^{(i)}\}$$

It should be noted that the stresses and strains are identified with their amplitude values in a harmonic loading cycle. All matrices in Equation (8.25) including the elasto-dissipative ones $-[\varphi]^{(k)}$, $[\chi]^{(k)}$, $[\psi]^{(k)}$ – are completely full.

Energy losses in a composite laminate are defined by EDC matrices

$$\Delta W = \frac{1}{2}\{\epsilon\}^{\mathrm{T}}[\varphi^{c}]\{\epsilon\}$$

$$\Delta W = \frac{1}{2}\{\epsilon\}^{\mathrm{T}}[\chi^{c}]\{\sigma\} \qquad (8.26)$$

$$\Delta W = \frac{1}{2}\{\sigma\}^{\mathrm{T}}[\psi^{c}]\{\sigma\}$$

where $\{\sigma\}$ and $\{\epsilon\}$ are average stresses and strains.

Relationships between the EDC matrices of the laminate and the EDC matrices of the layers can be determined by turning to strains in Equations (8.25) and (8.26) and comparing the corresponding relationships. Monolayer stresses and average laminate stresses are determined from strains with the help of the monolayer stiffness matrix $[G]^{(k)}$ and the laminate stiffness matrix $[G]$

$$\{\sigma\}^{(k)} = [G]^{(k)}\{\epsilon\}^{(k)}$$

$$\{\sigma\} = [G]\{\epsilon\} \qquad (8.27)$$

where the expression for $[G]$, considering Equation (8.4), is given by

$$[G] = \sum_{k=1}^{n} [G]^{(k)}\bar{h}^{(k)} \qquad (8.28)$$

Stiffness matrices $[G]^{(k)}$ in the (x_1, x_2) coordinate system are related to stiffness matrices $[G^0]^{(k)}$ in the $(x_1^{(k)}, x_2^{(k)})$ natural axes in the form:

$$[G]^{(k)} = [T_1^{(k)}][G^0]^{(k)}[T_1^{(k)}]^{\mathrm{T}} \qquad (8.29)$$

Here the transformation matrix $[T_1^{(k)}]$ is detailed in Equation (4.33), and the stiffness matrix of the monolayer $[G^0]$ is defined by its engineering constants from Equation (4.30).

Substituting Equation (8.27) into Equations (8.25) and (8.26) with consideration of Equation (8.1), one can derive the relationships for the laminate EDC matrices

$$[\varphi] = \sum_{k=1}^{n} \bar{h}^{(k)}[\varphi]^{(k)}$$

$$[\chi] = \left(\sum_{k=1}^{n} \bar{h}^{(k)}[\chi]^{(k)}[G]^{(k)} \right)[S] \qquad (8.30)$$

$$[\psi] = [S] \left(\sum_{k=1}^{n} \bar{h}^{(k)}[G]^{(k)}[\psi]^{(k)}[G]^{(k)} \right)[S]$$

where $[S]$ is the laminate compliance matrix $[S] = [G]^{-1}$.

Matrix Equations (8.26) and (8.30) enable one to calculate energy losses during the loading cycle from the mean stresses and/or strains for a multilayered composite with an arbitrary reinforced structure. Equations (8.30) are rather cumbersome in detailed form, but for special cases of the laminate structure they can be simplified considerably.

8.2.1 Angle-Plied Composites

Let us assume that all monolayers in the composite have the same stiffness and EDC strain matrices in the natural coordinate system:

$$[G^0] = \begin{bmatrix} g^0_{11} & g^0_{12} & 0 \\ g^0_{12} & g^0_{22} & 0 \\ 0 & 0 & g^0_{66} \end{bmatrix}, \qquad [\varphi^0] = \begin{bmatrix} \varphi^0_{11} & \varphi^0_{12} & 0 \\ \varphi^0_{12} & \varphi^0_{22} & 0 \\ 0 & 0 & \varphi^0_{66} \end{bmatrix} \qquad (8.31)$$

Angle-plied composites are orthotropic materials if they consist of an even number of monolayers, half of which are arranged at an angle, $+\alpha$ to the x_1-axis and the rest of which are arranged at an angle, $-\alpha$ (see Figure 8.5). Such orthotropic materials have matrices $[\varphi]$ and $[G]$ of the form

$$[Y] = \begin{bmatrix} y_{11} & y_{12} & 0 \\ y_{12} & y_{22} & 0 \\ 0 & 0 & y_{66} \end{bmatrix} \qquad (8.32)$$

where y is changed to g^0 for the $[G]$ matrix and to φ for the matrix $[\varphi]$.

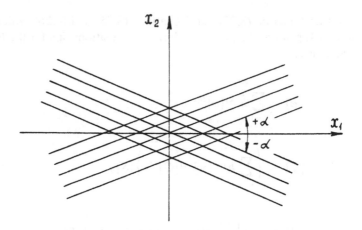

FIGURE 8.5. The pile-up of monolayers in the angle-plied composite.

Let us determine the specific form of the EDC stress matrix $[\psi]$. Using the transformation formulas (which are analogous to Equation (4.103) for the components of the monolayer stiffness matrices $[G^+]$ and $[G^-]$) and the transformation Equation (4.101) for the components of the monolayer EDC stress matrices $[\psi^+]$ and $[\psi^-]$ (signs $+$ and $-$ correspond to reinforcing angles $+\alpha$ and $-\alpha$), one can see that the pairs of matrices differ only in the signs before the components with subscripts 16 and 26

$$[G^{\pm}] = \begin{bmatrix} g_{11} & g_{12} & \pm g_{16} \\ g_{12} & g_{22} & \pm g_{26} \\ \pm g_{16} & \pm g_{26} & g_{66} \end{bmatrix}, \quad [\psi^{\pm}] = \begin{bmatrix} \psi_{11} & \psi_{12} & \pm \psi_{16} \\ \psi_{12} & \psi_{22} & \pm \psi_{26} \\ \pm \psi_{16} & \pm \psi_{26} & \psi_{66} \end{bmatrix}$$

$$(8.33)$$

Introducing auxiliary matrices produces

$$[G]^{(+)} = \begin{bmatrix} 0 & 0 & g_{16} \\ 0 & 0 & g_{26} \\ g_{16} & g_{26} & 0 \end{bmatrix}, \quad [\psi]^{(0)} = \begin{bmatrix} \psi_{11} & \psi_{12} & 0 \\ \psi_{12} & \psi_{22} & 0 \\ 0 & 0 & \psi_{66} \end{bmatrix}$$

$$[\psi]^{(+)} = \begin{bmatrix} 0 & 0 & \psi_{16} \\ 0 & 0 & \psi_{26} \\ \psi_{16} & \psi_{26} & 0 \end{bmatrix} \quad (8.34)$$

Then let us represent the matrices of Equation (8.33) as the sums

$$[G^{\pm}] = [G] \pm [G]^{(+)}, \quad [\psi^{\pm}] = [\psi]^{(0)} \pm [\psi]^{(+)} \quad (8.35)$$

Using Equations (8.30), (8.33)–(8.35) one can get

$$[\psi] = [\psi]^{(0)} + [G]^{-1}[P][G]^{-1} \qquad (8.36)$$

where the matrix $[P]$ is of the form

$$[P] = \begin{bmatrix} p_{11} & p_{12} & 0 \\ p_{12} & p_{22} & 0 \\ 0 & 0 & p_{66} \end{bmatrix}$$

and its components are written as

$$p_{11} = g_{16}(2\psi_{16}g_{11} + 2\psi_{26}g_{12} + \psi_{66}g_{16})$$

$$p_{12} = \psi_{16}(g_{26}g_{11} + g_{12}g_{16}) + \psi_{26}(g_{22}g_{16} + g_{26}g_{12}) + \psi_{66}g_{16}g_{26}$$
$$(8.37)$$

$$p_{22} = g_{26}(2\psi_{16}g_{12} + 2\psi_{26}g_{22} + \psi_{66}g_{26})$$

$$p_{66} = \psi_{11}g_{16}^2 + 2\psi_{12}g_{16}g_{26} + 2\psi_{16}g_{16}g_{66} + 2\psi_{26}g_{26}g_{66} + \psi_{22}g_{26}^2$$

Energy losses of the angle-plied composite during the loading cycle under an arbitrary plane stress state [Equation (8.26)] are determined with the help of the EDC matrix $[\psi]$ [Equation (8.36)]. The amplitude of the elastic energy W in the loading cycle is characterized by the compliance matrix $[S]$

$$W = \frac{1}{2}\{\sigma\}^T [S]\{\sigma\} \qquad (8.38)$$

where $[S]$ has the following components

$$[S] = [G]^{-1} = \frac{1}{g_{11}g_{22} - g_{12}^2} \begin{bmatrix} g_{22} & -g_{12} & 0 \\ -g_{12} & g_{11} & 0 \\ 0 & 0 & \dfrac{g_{11}g_{22} - g_{12}^2}{g_{66}} \end{bmatrix}$$

$$(8.39)$$

Let the angle-plied composite be under uniaxial cyclic loading along the x_1-axis ($\sigma_1 \neq 0$, $\sigma_2 = 0$, $\tau_{12} = 0$). In this case the energy losses, ΔW

[Equation (8.26)], and the amplitude value of the elastic energy, W [Equation (8.38)], are equal to

$$\Delta W = \frac{1}{2}\psi_{11}\sigma_1^2, \qquad W = \frac{1}{2}s_{11}\sigma_1^2$$

and the corresponding dissipation factor ψ looks as follows:

$$\psi = \Delta W/W = \psi_{11}/s_{11} \tag{8.40}$$

The component ψ_{11} of the matrix $[\psi]$ is defined from Equation (8.36). Multiplying the matrix $[P]$ from the left and from the right by the compliance matrix [Equation (8.39)] and then, considering the expressions for matrix $[P]$ components [Equation (8.37)] and $[S]$ [Equation (8.39)], one can derive

$$\psi = \frac{\psi_{11}(g_{11}g_{22} - g_{12}^2)}{g_{22}} + \frac{2\psi_{16}(g_{16}g_{22} - g_{26}g_{12})}{g_{22}}$$

$$+ \frac{\psi_{66}(g_{16}g_{22} - g_{26}g_{12})^2}{g_{22}(g_{11}g_{22} - g_{12}^2)} \tag{8.41}$$

The components ψ_{11}, ψ_{16}, and ψ_{66} of the EDC stress matrix, as well as the components of the monolayer stiffness matrix g_{11}, g_{12}, g_{22}, g_{16}, and g_{66} are related to the components of the correspondence matrices in the natural axes. Figures 8.6–8.8 plot the dependencies of dissipation factors and Young's moduli on the angle α for angle-plied composites under uniaxial loading along the x_1-axis. Initial data for calculation are given in Chapter 3. The three-component model for monolayer EDC was used when calculating the diagrams. Young's moduli E_1 that are inverse to the compliance matrix components with subscript 11 are determined by Equations (8.8). Figures 8.6–8.8 illustrate good agreement between predicted and experimental data.

8.3 ENERGY DISSIPATION OF MULTILAYERED COMPOSITES UNDER BENDING

Bending is a rather typical kind of loading for composite materials. Let us derive the relation for the energy dissipation in the multilayered composite plate in terms of the monolayer EDC (i.e., the monolayer elasto-dissipative characteristics).

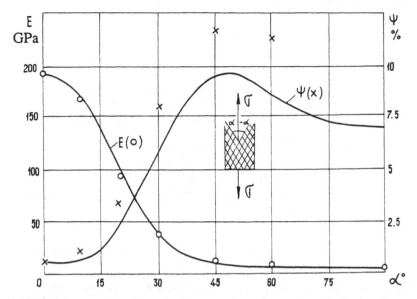

FIGURE 8.6. Dissipation factor, ψ, and Young's modulus, E, of angle-plied carbon fiber reinforced plastic (CFRP) HMS/DX 209 under uniaxial loading versus fiber orientation angle, α. Experimental data are shown by the points [2], predicted diagrams are shown by full lines [Equation (8.41)].

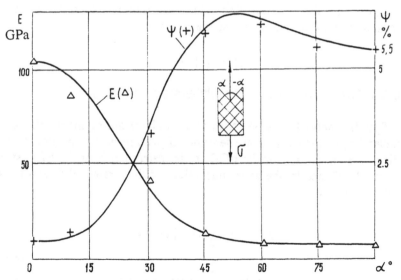

FIGURE 8.7. Dissipation factor, ψ, and Young's modulus, E, of angle-plied carbon fiber reinforced plastic (CFRP) HTS/DX 210 under uniaxial loading versus fiber orientation angle, α. Experimental data are shown by the points [2], predicted diagrams are shown by full lines [Equation (8.41)].

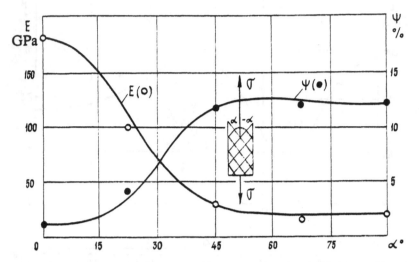

FIGURE 8.8. Dissipation factor, ψ, and Young's modulus, E, of angle-plied boron fiber reinforced plastic (BFRP) under uniaxial loading versus fiber orientation angle, α. Experimental data are shown by the points [25], predicted diagrams are shown by full lines [Equation (8.41)].

Let us suppose (as was done during the analysis of elastic characteristics) that the layers are bonded ideally, and the Kirchhoff-Love hypothesis from classical plate theory is fulfilled. The hypothesis results in the following relationships for the strains in the matrix form:

$$\{\epsilon\} = \{e\} + z\{\varkappa\}$$

8.3.1 The General Case of Multilayered Laminate

Specific energy losses $\Delta W^{(k)}$ in each kth layer during the loading cycle under the plane stress state are determined [with respect to the global coordinate system (x_1, x_2)] by the EDC strain matrix $[\varphi]^{(k)}$ or the EDC stress matrix $[\psi]^{(k)}$ and by the amplitude values of the strains $\{\epsilon\}^{(k)}$ or stresses $\{\sigma\}^{(k)}$ as follows:

$$\Delta W^{(k)} = \frac{1}{2}\{\epsilon\}^{(k)}[\varphi]^{(k)}\{\epsilon\}^{(k)}$$

$$\Delta W^{(k)} = \frac{1}{2}\{\sigma\}^{(k)}[\psi]^{(k)}\{\sigma\}^{(k)}$$

(8.42)

The components of the matrices $[\varphi]^{(k)}$ and $[\psi]^{(k)}$ are expressed in terms of the components of corresponding elasto-dissipative matrices in the natural coordinate system $(x_1^{(k)}, x_2^{(k)})$ from transformation formulas. Stress matrices $\{\sigma\}^{(k)}$, strain matrices $\{\epsilon\}^{(k)}$, as well as elasto-dissipative matrices $[\varphi]^{(k)}$ and $[\psi]^{(k)}$ are related to each other with the help of the monolayer stiffness matrix $[G]^{(k)}$:

$$\{\sigma\}^{(k)} = [G]^{(k)}\{\epsilon\}^{(k)}; \qquad [\varphi]^{(k)} = [G]^{(k)}[\psi]^{(k)}[G]^{(k)} \qquad (8.43)$$

Let us determine energy losses ΔW in the multilayered composite with unit dimensions along the (x_1, x_2) axes and with a height, H, corresponding to the laminate thickness. Now perform the integration with respect to the height of each layer in the first Equation (8.42), taking into account Equation (8.10), and then sum the energy losses in all the monolayers. The following equation is derived in the matrix form:

$$\Delta W = \frac{1}{2}(\{e\}^T[\varphi_1]\{e\} + 2\{e\}^T[\varphi_2]\{x\} + \{x\}^T[\varphi_3]\{x\}) \qquad (8.44)$$

where the matrices $[\varphi_1]$, $[\varphi_2]$, and $[\varphi_3]$ are related to EDC matrices in the form

$$[\varphi_1] = \sum_{k=1}^{n} \int_{z^{(k-1)}}^{z^{(k)}} [\varphi]^{(k)}dz$$

$$[\varphi_2] = \sum_{k=1}^{n} \int_{z^{(k-1)}}^{z^{(k)}} [\varphi]^{(k)}z\,dz \qquad (8.45)$$

$$[\varphi_3] = \sum_{k=1}^{n} \int_{z^{(k-1)}}^{z^{(k)}} [\varphi]^{(k)}z^2dz$$

Equations (8.45) can be written in the reduced form

$$[\varphi_j] = \sum_{k=1}^{n} \int_{z^{(k-1)}}^{z^{(k)}} [\varphi]^{(k)}z^{j-1}dz \qquad (j = 1,2,3) \qquad (8.46)$$

If the components of the matrices $[\varphi]^{(k)}$ do not vary with the thickness of the corresponding monolayers, then Equations (8.46) take the simpler form

$$[\varphi_j] = \frac{1}{j} \sum_{k=1}^{n} [\varphi]^{(k)} (z^j_{(k)} - z^j_{(k-1)})$$

where $j = 1,2,3$; $k = 1,2,\ldots,n$.

Once the auxiliary matrices $[\varphi_j]$ ($j = 1,2,3$) are combined, one can write Equation (8.44) in the detailed matrix form

$$\Delta W = \frac{1}{2} \begin{Bmatrix} e \\ x \end{Bmatrix}^{T} \begin{bmatrix} [\varphi_1] & [\varphi_2] \\ [\varphi_2] & [\varphi_3] \end{bmatrix} \begin{Bmatrix} e \\ x \end{Bmatrix} \qquad (8.47)$$

The matrices $[\varphi_j]$ ($j = 1,2,3$), $[\varphi]^{(k)}$, $[\psi]^{(k)}$ and $[G]^{(k)}$ ($k = 1,2,\ldots, n$) in Equations (8.42)–(8.47) are symmetric. Generally speaking, for a laminate with arbitrary pile-up of monolayers, these matrices are completely full. The matrices $[\varphi]^{(k)}$, $[\psi]^{(k)}$, and $[G]^{(k)}$ are of the simplest form in the natural coordinate system of the kth layer $(x_1^{(k)}, x_2^{(k)})$.

Energy losses ΔW in the multilayered composite during the loading cycle can be represented as a function of the load factors, Equation (8.12). Energy losses in composites under cyclic loading are governed by the amplitude values of the loading parameters. Load factors and strains are represented by their amplitude values in the loading cycle. Substituting Equation (8.25) into Equation (8.47), one obtains

$$\Delta W = \frac{1}{2} \begin{Bmatrix} T \\ M \end{Bmatrix}^{T} \begin{bmatrix} [S_1] & [S_2] \\ [S_2] & [S_3] \end{bmatrix} \begin{bmatrix} [\varphi_1] & [\varphi_2] \\ [\varphi_2] & [\varphi_3] \end{bmatrix} \begin{bmatrix} [S_1] & [S_2] \\ [S_2] & [S_3] \end{bmatrix} \begin{Bmatrix} T \\ M \end{Bmatrix} \qquad (8.48)$$

Note that to calculate the amplitude value of the potential energy of a multilayered composite, it is sufficient to use one of the Equations (8.44), (8.47), or (8.48), replacing φ by g and the notation ΔW by W.

8.3.2 Load Modes of Multilayered Composites. The Mode of Free Bending

Equation (8.48) is true for the general case of multilayered composite loading under a full six-component combination of forces $\{T\}$ and moments $\{M\}$. It is obvious that a considerable number of specific load modes can be realized in practice. The exact determination of the load mode is of particular importance for the analysis of experimental data.

Let us consider some specific cases of loading. Assume that the following components of the load vector are given:

$$M_1 \neq 0, \qquad M_2 = 0, \qquad M_{12} = 0, \qquad \{T\} = \{0\} \qquad (8.49)$$

It follows from Equation (8.25) that in this case not only does the curvature $\varkappa_1$ differ from zero, but $\varkappa_2 \neq 0$ and $\varkappa_{12} \neq 0$, as well. In the general case, the coordinate plane is also deformed

$$e_1 \neq 0, \qquad e_2 \neq 0, \qquad e_{12} \neq 0$$

This means that the load mode according to Equation (8.49) can be realized when the strains of the multilayered composite are not constrained. Let us call such a load mode the mode of free bending. The most widespread way to determine composite damping characteristics is to test flat prismatic specimens under bending vibrations. Dissipation factors of the specimens under bending vibrations are usually believed to be equal to dissipation factors under plane uniaxial loading of the specimens, i.e., in the load mode

$$T_1 \neq 0, \qquad T_2 = 0, \qquad T_{12} = 0, \qquad \{M\} = \{0\} \qquad (8.50)$$

In particular, all experimental data in Figures 8.6–8.8 were obtained assuming that damping characteristics under cyclic bending are identical to the characteristics under loading in a plane.

Equations (8.48) indicate that in the general case such an identity is not true. (Further, it will be shown that composite structures with the properties given in Figures 8.6–8.8 are pleasant exceptions to this case, and such an identity is quite reasonable for them.)

Specimen fastening during the test, as a rule, meets the condition of Equation (8.49) approximately. It is possible to get a sense of the influence of strain constraints in bending on the dissipation factor magnitude, if we consider two more load modes besides the mode of free bending.

The first mode can be called the mode of bending load with constrained torsional strains:

$$M_1 \neq 0, \qquad M_2 = 0, \qquad \varkappa_{12} = 0 \ (M_{12} \neq 0), \qquad \{\Gamma\} = \{0\}$$
$$(8.51)$$

From Equation (8.25), according to the value of M_1 and considering the condition $\varkappa_{12} = 0$, one can determine the corresponding value of M_{12}. Energy losses can be determined from Equation (8.48). They are not equal, in the general case, to the energy dissipation in the material under the free bending mode.

For the second example, let us consider the mode of bending by the moment M_1 when bending along the x_2-axis is constrained:

$$M_1 \neq 0, \quad \varkappa_2 = 0 \ (M_2 \neq 0), \quad M_{12} = 0, \quad \{T\} = \{0\} \quad (8.52)$$

In this case, considering the value of M_1 and the condition $\varkappa_2 = 0$, the value of the moment M_2 can be found from Equation (8.25). Energy losses can be calculated from Equation (8.48). They do not coincide, in the general case, with energy losses when loading the composite in the modes of Equations (8.49) and (8.51).

Comparison of dissipation factor values for specific types of composites under the load modes of Equations (8.49), (8.50), (8.51), and (8.52) enables one to draw conclusions about the effect of strain constraints during specimen tests. Figure 8.9 represents the dissipation factors of undirectional composite for different kinds of specimen fastening. As one can see, the difference in dissipation factor values—which correspond to different fastening conditions—can be very perceptible. Unfortunately, it should be noted

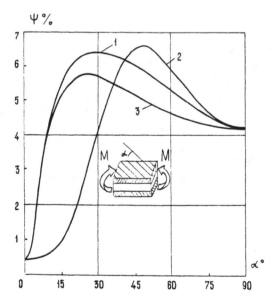

FIGURE 8.9. Dissipation factor, ψ, of unidirectional carbon fiber reinforced plastic (CFRP) HMS/DX 210 versus the angle, α, between the fiber direction and the bending plane: 1—the mode of free bending; 2—Equation (8.51) (constrained torsion); and 3—Equation (8.52) (constrained bending in the plane orthogonal to the bending plane).

that the authors of a lot of experimental works omit the details of specimen fastening in their experiments as insignificant details. This makes it very difficult to compare the results of different authors.

8.3.3 Symmetrical Structures of Multilayered Hybrid Composites. Specific Composite Types

Equation (8.48) becomes simpler for a multilayered composite when pile-up of the layers is symmetric with respect to the mid-plane, selected as the coordinate plane:

$$[S_1] = [G_1]^{-1}, \qquad [S_2] = 0, \qquad [S_3] = [G_3]^{-1} \qquad (8.53)$$

$$\Delta W = \frac{1}{2} \begin{Bmatrix} T \\ M \end{Bmatrix}^{\mathsf{T}} \begin{bmatrix} [S_1][\varphi_1][S_1] & 0 \\ 0 & [S_3][\varphi_3][S_3] \end{bmatrix} \begin{Bmatrix} T \\ M \end{Bmatrix} \qquad (8.54)$$

Equations (8.54) and (8.24) enable one to calculate energy losses and the potential energy for a plate constituted of different monolayers with arbitrary orientation (hybrid composite) under the action of the forces and the moments. To do this, it is sufficient to know the EDC matrices of each layer [Equations (8.42) and (8.43)], the geometrical parameters of the layers [see Equation (8.46)], and the exerted force factors.

Figures 8.10–8.15 illustrate the predicted dependencies which were calculated on the basis of Equation (8.50), as well as experimental data for several composites. The composites were loaded by the bending moment, provided that the plane of the bending moment action changed its orientation with respect to composite axes. The four-component model of the EDC was used for carbon fiber reinforced monolayers HMS/DX 210 and glass fiber reinforced monolayers GLASS/DX 210 (see Table 3.1). All calculations were performed for the mode of free bending. The results for orthogonally reinforced composites, as well as hybrid and symmetrical composites of the general structure, show good agreement between theoretical and experimental data.

Let us derive specific forms for the matrices $[\varphi_j]$, $[G_j]$ $(j = 1,3)$ for special composite types with symmetrical pile-up of n similar monolayers in the plate with the total thickness, H. The elasto-dissipative behavior of the monolayer in the general coordinate system (x_1, x_2) is determined by the stiffness $[G]$ and the compliance $[S]$ matrices $([S] = [G]^{-1})$, and the EDC strain or stress matrices, $[\varphi]$ or $[\psi]$. Their components can be calculated from the components of these matrices in the natural monolayer axes and can be expressed in terms of engineering elasto-dissipative constants from equations of the (4.101)–(4.103) type.

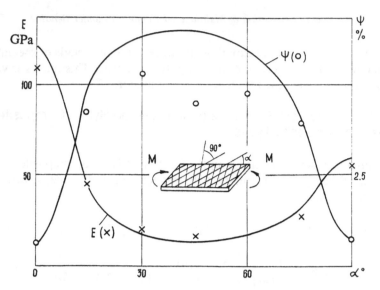

FIGURE 8.10. Dissipation factor, ψ, and effective Young's modulus, E, of the cross-plied CFRP HMS/DX 210 versus fiber orientation angle, α, with respect to the plane of bending moment. Experimental data are shown by the points [52], predicted diagrams are drawn by full lines. Input data are listed in Chapter 3, Table 3.1. Laminate structure is as follows: $[0°,90°,0°,90°]_s$.

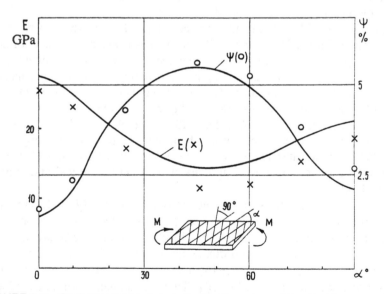

FIGURE 8.11. Dissipation factor, ψ, and effective Young's modulus, E, of the cross-plied GFRP GLASS/DX 210 versus fiber orientation angle, α, with respect to the plane of bending moment. Experimental data are shown by the points [52], predicted diagrams are shown by full lines. Input data are listed in Chapter 3, Table 3.1. Laminate structure is as follows: $[0°,90°,90°,0°,90°]_s$.

172

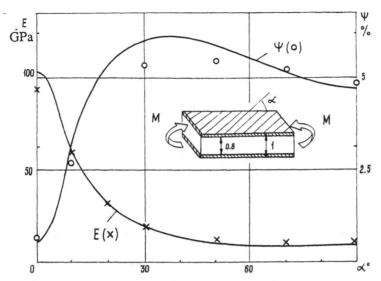

FIGURE 8.12. Dissipation factor, ψ, and effective Young's modulus, E, of the unidirectional hybrid composite versus fiber orientation angle, α, with respect to the plane of bending moment. Outer layers are made of CFRP HMX/DX 210, the inner layer is made of GFRP GLASS/DX 210. Relative layer thicknesses are given in the figure. Experimental data are shown by the points [54], predicted diagrams are drawn by full lines. Input data are listed in Chapter 3, Table 3.1.

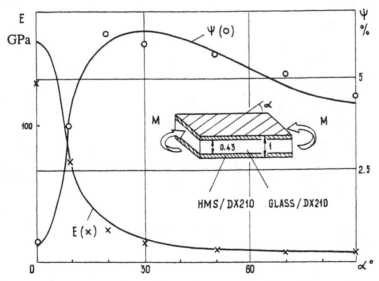

FIGURE 8.13. Dissipation factor, ψ, and effective Young's modulus, E, of the unidirectional hybrid carbon/glass fiber reinforced plastic versus fiber orientation angle, α, with respect to the plane of bending moment. Outer layers are made of CFRP HMS/DX 210, the inner layer is made of GFRP GLASS/DX 210. Relative layer thicknesses are given in the figure. Experimental data are shown by the points [54], predicted diagrams are drawn by full lines. Input data are listed in Chapter 3, Table 3.1.

173

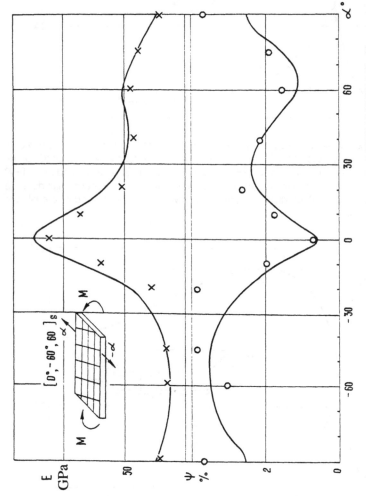

FIGURE 8.14. Dissipation factor, ψ, and effective Young's modulus, E, of CFRP HMS/DX 210 with symmetrical pile-up of the layers [0°, −60°, 60°]$_s$ versus angle, α (the angle between the fiber direction in the 0° layer and the bending moment plane). Experimental data are shown by the points [52], predicted diagrams are drawn by full lines.

174

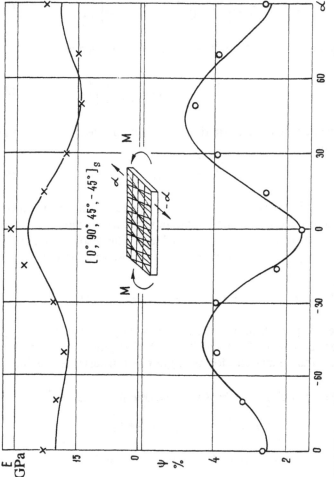

FIGURE 8.15. Dissipation factor, ψ, and effective Young's modulus, E, of GFRP GLASS/DX 210 with symmetrical pile-up of the layers $[0°, 90°, 45°, -45°]_s$ versus angle, α (the angle between the fiber direction in the $0°$ layer and the bending moment plane). Experimental data are shown by the points [52], predicted diagrams are drawn by full lines.

175

Let us consider two types of composites. The first one is a unidirectional material with an orientation angle, α, to the x_1-axis. The second material is an angle-plied material ($\pm\alpha$).

The relationship for energy losses in the unidirectional material, using Equations (8.46) and (8.54), is expressed as

$$\Delta W = \frac{1}{2}\begin{Bmatrix} T \\ M \end{Bmatrix}^{\mathrm{T}} \begin{bmatrix} [\psi]/H & 0 \\ 0 & 12[\psi]/H^3 \end{bmatrix} \begin{Bmatrix} T \\ M \end{Bmatrix} \qquad (8.55)$$

The amplitude value of the potential energy is defined as:

$$W = \frac{1}{2}\begin{Bmatrix} T \\ M \end{Bmatrix}^{\mathrm{T}} \begin{bmatrix} [S]/H & 0 \\ 0 & 12[S]/H^3 \end{bmatrix} \begin{Bmatrix} T \\ M \end{Bmatrix} \qquad (8.56)$$

The dissipation factor of the unidirectional composite under free bending is of the form

$$\psi = \psi_{11}/s_{11} \qquad (8.57)$$

Equation (8.57) coincides with the expression for the dissipation factor in the case of uniaxial cyclic loading of the unidirectional material at an angle to the reinforcing direction [Equation (4.105)]. This also can be concluded directly from Equations (8.55) and (8.56), considering the following load mode:

$$T_1 \neq 0, \qquad T_2 = 0, \qquad T_{12} = 0, \qquad \{M\} = \{0\}$$

For the angle-plied material, the matrices $[G_j]$ and $[\varphi_j]$ ($j = 1,3$) are defined from Equations (8.17) and (8.46). Uniting the expressions for the components of these matrices for an angle-plied material results in:

$$(y_{11j}, y_{12j}, y_{22j}, y_{66j}) = a_j(y_{11}, y_{12}, y_{22}, y_{66}), \qquad (j = 1,3)$$

$$a_1 = H, \qquad a_1 = H^3/12$$

$$(y_{16j}, y_{26j}) = b_j(y_{16}, y_{26}), \qquad (j = 1,3)$$

- if n is even, then the coefficients b_j equal: $b_1 = 0$, $b_3 = H^3/4n$
- if n is odd, then $b_1 = H/n$, $b_3 = (3n^2 - 2)H^3/4n^3$

where y must be changed to φ for the matrices $[\varphi_j]$ and to g for the matrices $[G_j]$; the components y_{16j}, y_{26j} ($j = 1,3$) have a sign that corresponds to the sign of the reinforcing angle, α, of the outer layers.

For the angle-plied composite the components of the matrices $[\varphi_j]$ and

$[G_j]$ ($j = 1,3$) with the subscripts 16 and 26 have the order of $1/n$ (n is the number of monolayers). If the number of composite monolayers is rather large, these coefficients are assumed to be zero. In the mode of free bending, Equations (8.54) and (8.57) give the following formula for the dissipation factor

$$\psi = \varphi_{11_3} s_{11_3} + 2\varphi_{12_3} s_{12_3} + \varphi_{22_3}(s_{12_3})^2/s_{11_3}$$

Using the relationship between the matrices $[\varphi]$ and $[\psi]$, the formula can be written in the form

$$\psi = \psi_{11}(g_{11}g_{22} - g_{12}^2)/g_{22} + 2\psi_{16}(g_{16}g_{22} - g_{26}g_{12})/g_{22}$$

$$+ \psi_{66}(g_{16}g_{22} - g_{26}g_{12})^2/g_{22}(g_{11}g_{22} - g_{12}^2) \qquad (8.58)$$

Equation (8.58) coincides with the formula for the dissipation factor of the angle-plied composite under uniaxial loading, Equation (8.41).

Using Figure 8.16, it is possible to estimate the effect of the number of

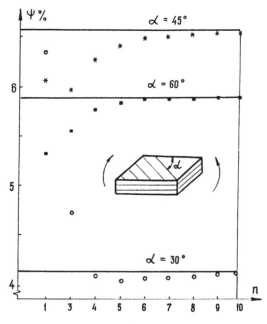

FIGURE 8.16. Effect of monolayers number, n of the angle-plied composite HMS/DX 210 with symmetrical pile-up of the layers on dissipation factor value, ψ, in the mode of free bending. The signs correspond to the following reinforcing angles: $\bigcirc$—±30°, *—±45°, ■—±60°; straight lines are asymptotes for $n \to \infty$. The values of ψ for unidirectional composite ($n = 1$) are given for comparison.

angle-plied composite monolayers on the values of the dissipation factors in the mode of free bending. Input data for calculation here are the same as for Figure 8.9. Equations (8.54) and (8.57) were used in calculating the values of $\psi(n)$. These equations result in Equation (8.58), when $n \rightarrow \infty$.

Figure 8.16 shows that if the number of layers in the angle-plied material equals 8 to 10, one can use the simple Equation (8.58). On the contrary, for a composite that consists of only three to five layers, the error can be very appreciable during calculation of the dissipation factor from this formula.

In conclusion it should be particularly emphasized that the above-mentioned coincidence of dissipation factors for unidirectional or angle-plied composites in free bending and uniaxial loading does not apply—in the general case—to composites of an arbitrary structure.

CHAPTER 9

Energy Dissipation in Vibrating Composite Bars

Energy losses in a unit volume of material during a loading cycle under different kinds of stress states were described in preceding chapters. The aim of this chapter is to apply the approximate energy method to the calculation of free damped vibrations in cantilever bars, taking into account the transverse shear.

The application of the energy method is directly related to the analysis of the energy balance equation, considering energy losses for the internal friction. To calculate the system energy, it is necessary to know the modes and the frequencies of the vibrations. When the losses for internal friction are small, the modes and the frequencies of damped and sustained vibrations differ slightly. This is confirmed by predicted and experimental results. That is why it is necessary to find approximate values of the terms in the energy balance equation from the modes and the frequencies of perfectly elastic vibrations.

The energy method can be used successfully for a variety of dynamic problems involving vibrations [57]. In particular, Reference [15] considers the simplest mechanical systems. It was shown that in the case of viscous internal friction, the energy method and the exact solution of differential equations produce identical values for the amplitudes of free damped and constrained vibrations. In the case where frictional forces are proportional to the displacement (the strain), the values of these amplitudes differ slightly.

9.1 FREE DAMPED VIBRATIONS OF CANTILEVER BARS. EQUATIONS OF MOTION

Let us consider free monoharmonical vibrations of the cantilever bar. Write the energy balance equation for two neighboring positions of the system, when the displacements reach amplitudes A and A' (see Chapter 2, Figure 2.3). The velocity of the system points reduces to zero at the extreme positions. The kinetic energy equals zero here and the energy balance equation reduces to the variation of the system potential energy due to energy losses in the cycle of vibrations

$$\Pi(A') = \Pi(A) - \Delta\Pi\left(\frac{A + A'}{2}\right) \tag{9.1}$$

Here $\Pi(A)$ is the amplitude of the potential energy for the displacement amplitude A; while $\Delta\Pi(A + A')/2$ is the energy loss in the cycle of vibrations, corresponding to the mean amplitude value of the displacement $(A + A')/2$. To calculate the terms in Equation (9.1), it is necessary to know the frequencies and the modes of (the) vibrations for the elastic system.

An infinite range of modes and frequencies arises following arbitrary initial excitation of free elastic vibrations. The general solution can be represented as the sum of individual monoharmonical vibrations. The exact solution corresponds to the infinite number of the terms. If the system has internal friction, then, as a rule, not long after the beginning of the free damped motion the system will execute the vibrations according to one of the lowest modes. The highest modes of the vibrations will damp quickly due to improved energy dissipation. Equation (9.1) can be used to describe the stage of monoharmonical motion.

Let us consider a cantilever bar with two types of cross sections and reinforced structures (see Figure 9.1). The bar in Figure 9.1(a) is a multilayered

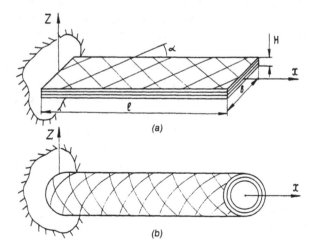

FIGURE 9.1. Structure and forms of laminated cantilever bars.

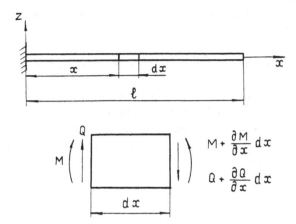

FIGURE 9.2. For derivation of equation for vibrating cantilever bar with consideration for shear.

laminate that is symmetrical with respect to its mid-plane. It consists of angle-plied layers. The bar can also be constituted from layers of an orthotropic material, whose main axes of anisotropy coincide with the axes x, y, and z of the bar. Assume that the number of pairs of monolayers in the angle-plied composite is sufficiently large.

The round hollow bar in Figure 9.1(b) has a cylindrical symmetrical laminated structure, consisting of pairs of angle-plied composite layers. It can also be built up from layers of cylindrically orthotropic material, where each neighboring pair of angle-plied monolayers is considered to be a single orthotropic layer.

The bars can execute bending vibrations in the (z, x) plane. To determine the frequencies and the modes of vibrations, we will set up an equation for the vibrations of the cantilever bar (Figure 9.2), restricting our consideration to shear strains and applying the scheme proposed by S. P. Timoshenko [82].

The angle between the tangent to the curved mid-axis of the deformed bar and the x-axis is determined as

$$\frac{\partial z}{\partial x} = \eta + \gamma_0 \qquad (9.2)$$

where η is the angle of slope of the tangent to the mid-axis, when one neglects shear, and γ_0 is the shear angle at the bar mid-plane level.

The moment, M, and the transverse force, Q, are related to the strains by the following relationships:

$$M = Dx = D\frac{\partial\eta}{\partial x}, \quad Q = k'\gamma_0 = k'\left(\frac{\partial z}{\partial x} - \eta\right) \tag{9.3}$$

where D is the bar bending stiffness, k' is a coefficient depending upon the form of the cross section, the material elastic parameters, and the law of shear strains distribution with the cross section height.

Let us now set up an equation for rotation and progressive motion of a unit bar fragment with a length dx (see Figure 9.2). As rotation inertia is neglected

$$Q dx - \frac{\partial M}{\partial x}dx = 0 \tag{9.4}$$

In line with Equation (9.3), Equation (9.4) takes the form:

$$D\frac{\partial^2\eta}{\partial x^2} = k'\left(\frac{\partial z}{\partial x} - \eta\right) \tag{9.5}$$

The equation for progressive motion of the unit bar fragment is as follows:

$$\frac{\partial Q}{\partial x}dx = \varrho\frac{\partial^2 z}{\partial t^2}dx \tag{9.6}$$

Here ϱ is the mass of the unit bar fragment. Performing transformation of Equations (9.5) and (9.6), it is possible to obtain an equation in the single function $z(x,t)$ for bar vibrations, considering shear

$$\frac{\partial^4 z}{\partial x^4} - a\frac{\partial^4 z}{\partial x^2\partial t^2} + b\frac{\partial^2 z}{\partial t^2} = 0 \tag{9.7}$$

where $a = \varrho/k', b = \varrho/D$.

Now it is necessary to set up boundary conditions for the function $z(x,t)$. The cantilever bar must meet the following requirements: one end plane is clamped rigidly and the second one is not subjected to the action of the moment and the transverse force, i.e.,

$$
\begin{array}{llll}
x = 0 & z = 0, & \partial z/\partial x = 0 & [2] \\
x = l & Q = 0 & M = 0 &
\end{array}
\tag{9.8}
$$

[2]More correct boundary conditions can be obtained by the variation method [see Alfutov, N. A., ed., 1991. *Principles of Stability Analysis for Elastic Systems* (in Russian). Moscow: Mashinostroenie]. Then the condition $dz/dx = 0$ should be changed for $\eta = 0$.

Considering Equations (9.3), (9.5), and (9.6), let us write boundary conditions with respect to the function $z(x,t)$ in the form:

$$\text{if } x = 0, \quad z = 0, \quad \partial z/\partial x = 0$$

$$\text{if } x = l, \quad -a\frac{\partial^3 z}{\partial x\,\partial t^2} + \frac{\partial^3 z}{\partial x^3} = 0, \quad -a\frac{\partial^2 z}{\partial t^2} + \frac{\partial^2 z}{\partial x^2} = 0$$

(9.9)

Let us now find the solution of Equation (9.7) with boundary conditions [Equation (9.9)] by the variable-separation method. Represent the function z as

$$z(x,t) = AX(x) \sin \omega t \qquad (9.10)$$

Substituting Equation (9.10) in Equation (9.7), one will obtain the linear uniform differential equation with respect to the function $X(x)$,

$$\frac{d^4 X}{dx^4} + a\omega^2\frac{d^2 X}{dx^2} - b\omega^2 X = 0 \qquad (9.11)$$

where the function $X(x)$ is called the normal function or the mode of vibrations corresponding to the frequency, ω.

The characteristic equation corresponds to Equation (9.11), as follows:

$$m^4 + a\omega^2 m^2 - b\omega^2 = 0$$

The characteristic equation has two real and two imaginary roots such that

$$m_1^2 = -\frac{a\omega^2}{2} + \sqrt{\frac{a^2\omega^4}{4} + b\omega^2}$$

(9.12)

$$m_2^2 = \frac{a\omega^2}{2} + \sqrt{\frac{a^2\omega^4}{4} + b\omega^2}$$

Here m_1 and m_2 are the modules of the roots of the equation, $a > 0, b > 0$.
The solution of Equation (9.11) is of the form:

$$X(x) = C_1 \sin m_2 x + C_2 \cos m_2 x + C_3 \text{ sh } m_1 x + C_4 \text{ ch } m_1 x \qquad (9.13)$$

For the function $X(x)$ [Equation (9.13)], boundary conditions are brought to the combined equations with respect to the coefficients C_1, C_2, C_3, and C_4:

$$C_1 m_2 + C_3 m_1 = 0, \qquad C_2 + C_4 = 0$$

$$C_1(-m_2^3 + m_2 \omega^2 a) \cos m_2 l + C_2(m_2^3 - m_2 \omega^2 a) \sin m_2 l$$

$$+ C_3(m_1^3 - m_1 \omega^2 a) \text{ ch } m_1 l + C_4(m_1^3 + m_1 \omega^2 a) \text{ sh } m_1 l = 0$$

$$C_1(-m_2^2 + a\omega^2) \sin m_2 l + C_2(-m_2^2 + a\omega^2) \cos m_2 l$$

$$+ C_3(m_1^2 + a\omega^2) \text{ sh } m_1 l + C_4(m_1^2 + a\omega^2) \text{ ch } m_1 l = 0$$

$$(9.14)$$

The first two expressions in Equations (9.14) give the direct relation between the pairs of coefficients

$$C_4 = -C_2, \quad C_3 = -C_1 m_2/m_1 \tag{9.15}$$

The last two expressions in Equations (9.14), considering Equation (9.15), can be transformed to the following:

$$-C_1 m_2(q \cos m_2 l + p \text{ ch } m_1 l) + C_2(q m_2 \sin m_2 l - p m_1 \text{ sh } m_1 l) = 0$$

$$(9.16)$$

$$C_1(q m_1 \sin m_2 l + p m_2 \text{ sh } m_1 l) + C_2 m_1(q \cos m_2 l + p \text{ ch } m_1 l) = 0$$

where $q = m_2^2 - \omega^2 a$ and $p = m_1^2 + \omega^2 a$.

In order for the combined Equations (9.14) to have a nonzero solution, the determinant of them, Δ, must be equal to zero, i.e.,

$$\Delta = m_1 m_2(q^2 + p^2) + pq[(m_2^2 - m_1^2) \text{ sh } m_1 l \sin m_2 l$$

$$+ 2 m_1 m_2 \text{ ch } m_1 l \cos m_2 l] = 0 \tag{9.17}$$

Considering Equations (9.12) and (9.17), frequency-response equation takes the form:

$$2 \text{ ch } m_1 l \cos m_2 l + \frac{a\omega \text{ sh } m_1 l \sin m_2 l}{\sqrt{b}} + \frac{a^2 \omega^2}{b} + 2 = 0 \tag{9.18}$$

Equation (9.18) is the transcendental equation with respect to the frequencies of free vibrations, ω. The parameters a and b [Equation (9.7)] depend on bending and shear stiffnesses of the bar, and the parameters m_1 and m_2 [Equation (9.15)] moreover, depend on the frequency, ω.

As all coefficients C_i of Equation (9.14) are related to the magnitude of the coefficient C_1, it is possible to set the value of C_1 equal to 1. The common factor will enter into Equation (9.10) as the amplitude of vibrations, A. Then, from Equations (9.15) and (9.16) one will obtain:

$$C_1 = 1, \qquad C_2 = \frac{q \cos m_2 l + p \text{ ch } m_1 l}{q \sin m_2 l - m_1 p \text{ sh } m_1 l / m_2}$$

$$(9.19)$$

$$C_3 = -\frac{m_2}{m_1}, \qquad C_4 = -C_2$$

Consequently, to calculate the modes and the frequencies of vibrations for the cantilever bar, considering transverse shear, it is necessary to solve Equation (9.18) and to use Equations (9.12), (9.13), and (9.19). In addition, the parameters ϱ, k', and D [Equation (9.3)] must be known.

Let us derive the relationships for the coefficient k' of the transverse force [Equation (9.3)] and for energy losses in the bars for vibration modes described by Equation (9.13).

Taking the hypothesis of plane sections ($\gamma_0 = 0$), one has the following equation of vibrations instead of Equation (9.7):

$$\frac{\partial^4 z}{\partial x^4} + \frac{\varrho}{D} \frac{\partial^2 z}{\partial t^2} = 0 \qquad (9.20)$$

Let us find the solution of Equation (9.20) for monoharmonical vibrations in the form:

$$z(x,t) = AX(x) \sin \omega t \qquad (9.21)$$

where $X(x)$ is the normal mode of vibrations with consideration of shear.

Substitution of Equation (9.21) into Equation (9.20) results in an expression for the mode of vibrations:

$$X^{IV} - m^4 X = 0$$

$$(9.22)$$

$$m^4 = \varrho \omega^2 / D$$

Boundary conditions for free harmonic vibrations in a cantilever bar of length l are given in the form

$$x = 0 \qquad X = 0, \qquad \frac{dX}{dx} = 0$$

$$x = 1 \qquad \frac{d^2X}{dx^2} = 0, \qquad \frac{d^3X}{dx^3} = 0 \tag{9.23}$$

Equation (9.22), taking into account Equation (9.23), has a solution of the following form:

$$X = (\sin mx - \text{sh } mx) + \beta(\cos mx - \text{ch } mx) \tag{9.24}$$

where $\beta = (\cos ml + \text{ch } ml)/(\sin ml - \text{sh } ml)$, while the values of the parameter, m, and the vibration frequencies are determined from the frequency-response equation

$$\text{ch } ml \cos ml = -1 \tag{9.25}$$

The first three roots of Equation (9.25) are equal respectively to 1.875, 4.694, and 7.855. Equations (9.24), (9.25), and (9.21) present the solution of the problem on free sustained vibrations of the cantilever bar.

9.2 PLANE LAMINATED BAR

Let us consider the case of a plane laminated bar [see Figure 9.1(a)]. Suppose that shear strains, γ, in the (x,z) plane vary with the bar thickness according to the parabola law [82]

$$\gamma = \gamma_0(1 - 4z^2/H^2) \tag{9.26}$$

Here the coordinate, z, is measured from the mid-plane, and H is the bar thickness.

Equation (9.26) is among the possible hypotheses for bending, considering transverse shear. Let us consider a unit volume cut from the bar by its coordinate planes (Figure 9.3). We will ignore the stresses that develop because of the differences in the Poisson's ratios of the bar layers. In such a case, the unit volume is believed to be under uniaxial loading along the x-axis due to the bending moment, and under shear in the (x,z) plane due to

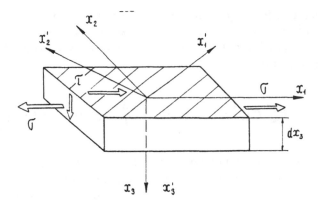

FIGURE 9.3. Stress state of a bar unit volume under bending with consideration for transverse shear. Coordinate systems (x,y,z) (Figure 9.2) and (x_1,x_2,x_3) for the bar coincide.

due to the transverse force. Normal stresses, σ, and shear stresses, τ, are related to the strains

$$\sigma = E_1\epsilon, \qquad \tau = G_{13}\gamma \qquad (9.27)$$

where E_1 is the Young's modulus of the orthotropic layer under its loading along the x-axis, G_{13} is the shear modulus in the (x,z) plane, and ϵ is the strain determined through the curvature, $\varkappa$ [Equations (9.3)], and the layer coordinate, z, with respect to the bar mid-plane, i.e., $\epsilon = \varkappa z$. From here on, a superscript which indicates that EDC values are related to certain bar layers will be omitted, except when the summation is over the layers. Each pair of monolayers is considered to be a united orthotropic layer. That is why the shear modulus G_{13} of such a layer is determined from the transformation of monolayer stiffness matrix components when rotating the coordinate system (x_1,x_2) about the z-axis (see Figure 9.3) [19]:

$$G_{13} = G'_{13} c^2 + G'_{23} s^2 \qquad (9.28)$$

Here G'_{13} and G'_{23} are the shear moduli of the monolayer in its natural coordinate system (x'_1,x'_2), s = sin α, c = cos α. Young's modulus, E_1, of the orthotropic material, when loading along the axis of symmetry of elastic properties, is related to the stiffness matrix components g_{ij} in the axes (x_1,x_2) as follows:

$$E_1 = g_{11} - g_{12}^2/g_{22} \qquad (9.29)$$

The transverse force, Q, in the cross section of the laminated bar [Figure 9.1(a)] is defined from the integral with respect to shear stresses, τ. Applying Equations (9.27) and (9.28), let us derive the expression for the coefficient k' in Equation (9.3)

$$k' = b\sum_{i=1}^{n} G_{13}^{(i)} \left\{ z^{(i)} - z^{(i-1)} - \frac{4}{3H^2}[(z^{(i)})^3 - (z^{(i-1)})^3] \right\} \quad (9.30)$$

Energy losses in the unit volume (Figure 9.3) can be represented as the sum of losses through uniaxial loading and shear. Thus, complete energy lost in the piece of the bar with unit length is equal to the sum of energy losses owing to the bending moment, ΔW_D, and to the transverse force, ΔW_s. The second item depends upon the action of shear stresses, τ, and is equal to

$$\Delta W_s = \frac{1}{2}b\sum_{i=1}^{n} \psi_{55}^{(i)} \int_{z^{(i-1)}}^{z^{(i)}} \tau^2 dz \quad (9.31)$$

Here $\psi_{55}^{(i)}$ is the component of the EDC stress matrix of the ith orthotropic layer [Equation (4.87)]. Its value is defined from the values of engineering elastic and dissipative constants [Equation (4.89)].

For angle-plied layers the component ψ_{55} is determined through the component Φ_{55} of the EDC strain matrix. The component Φ_{55} equals the component of the monolayer EDC matrix in the coordinate system which is rotated through an angle, α, with respect to the natural coordinate axes

$$\Phi_{55} = \Phi_{55}' c^2 + \Phi_{44}' s^2 \quad (9.32)$$

This formula is similar to Equation (9.28) for the components of the monolayer stiffness matrix. On the basis of the relationship between the matrices $[\Psi]$ and $[\Phi]$ [Equation (4.58)], $\psi_{55} = \Phi_{55}/G_{13}^2$, $\Phi_{55}' = \psi_{55}'G_{23}'^2$, and $\Phi_{44}' = \psi_{44}'G_{23}'^2$. Then the expression for the component Φ_{55} of the matrix $[\Phi]$ for the angle-plied layer is of the form:

$$\Phi_{55} = \psi_{55}'G_{13}'^2 c^2 + \psi_{44}'G_{23}'^2 s^2 \quad (9.33)$$

Here ψ_{55}' and ψ_{44}' are the components of the monolayer matrix in the natural axes (x_1', x_2', x_3'). Their values can be calculated from the engineering constants [Equation (4.89)] or else it is sufficient to assume, in accordance with the structural model [Equation (5.35)], that $\psi_{44}' = \psi_{55}' = \psi_{66}'$.

Considering Equations (9.26), (9.28), and (9.33), Equation (9.31) for energy losses takes the form:

$$\Delta W_s = \frac{1}{2} F_s \gamma_0^2 \qquad (9.34)$$

where

$$F_s = b \sum_{i=1}^{n} \Phi_{55}^{(i)} \left(z - \frac{8}{3} \frac{z^3}{H^2} + \frac{16}{5} \frac{z^5}{H^4} \right) \Bigg|_{z^{(i-1)}}^{z^{(i)}}$$

The expression for energy losses in the vibrating bar is of the following form:

$$\Delta \Pi = \Delta \Pi_D + \Delta \Pi_s \qquad (9.35)$$

$$\Delta \Pi_D = \frac{1}{2} F_D \int_0^l x^2 dx, \quad \Delta \Pi_s = \frac{1}{2} F_s \int_0^l \gamma_0^2 dx$$

When the hypothesis of plane sections is taken into account, then the relationships from Chapter 8, Section 8.3 may be applied to plane laminated bars. The bending moment and energy losses, ΔW_D, are referred to the unit length of the mid-plane (i.e., to the unit of the bar width b). The relation for the bar bending stiffness, D, can be derived, considering Equation (8.21), as follows:

$$D = b/s_{11}^3 \qquad (9.36)$$

where s_{11}^3 is the component of the compliance matrix [Equation (8.22)]. Energy losses, ΔW_D, and the potential energy, W_D [Equation (8.54)], in a volume with unit dimensions along the x- and y-axes and with a height that equals the bar height are of the following forms:

$$\Delta W_D = \frac{1}{2} [\varphi_{11}^3 (s_{11}^3)^2 + 2\varphi_{12}^3 s_{12}^3 s_{11}^3 + \varphi_{22}^3 (s_{12}^3)^2] M^2 = \frac{1}{2} \psi_{11}^3 M^2 \qquad (9.37)$$

$$W_D = \frac{1}{2} s_{11}^3 M^2$$

A new coefficient, ψ_{11}^3, is introduced here, which is the sum of the terms in square brackets.

The potential energy and energy losses under in vibrating bar, considering Equations (9.3), (9.36), and (9.37), can be calculated from the following formulas:

$$\Pi = \int_0^l bW_D dx = \frac{1}{2} D \int_0^l x^2 dx$$

$$\Delta\Pi = \int_0^l b\Delta W_D dx = \frac{1}{2} F_D \int_0^l x^2 dx$$

(9.38)

where $F_D = \psi_{11}^3 D^2/b$.

Taking into account the hypothesis of plane sections, the dissipation factor, ψ, does not depend on the modes of vibrations and equals the dissipation factor of the laminated material under ideal shear, i.e.,

$$\psi = \frac{\Delta\Pi}{\Pi} = \frac{\Delta W_D}{W_D} = \frac{\psi_{11}^3}{s_{11}^3}$$

(9.39)

Some results of calculations for cantilever plane bars with consideration of transverse shear strains are shown in Figures 9.4–9.6 and in Table 9.1. The EDC of carbon fiber reinforced plastic "Kulon" were used as input data (see Chapter 3).

Analysis of the data shown in Figures 9.4–9.6 and in Table 9.1 reveals that energy dissipation at the cost of transverse shear essentially depends on the bar elongation and the modes of vibrations. If reinforcing angles of the monolayers in angle-plied bars are small, then the energy dissipation at the cost of transverse shear can be considerably more than the dissipation due to the action of the bending moment.

It is common practice to ignore the corrections for vibration frequencies of the bars when using the Timoshenko model. The effect of transverse shear on the energy dissipation is much more significant and it must be specially analyzed in every specific case.

9.3 ROUND HOLLOW BAR

A round bar [Figure 9.1(b)] consists of layers which are made of a material with cylindrical orthotropy of elasto-dissipative properties. Let us

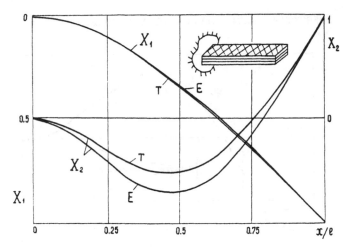

FIGURE 9.4. The first $[X_1 (x/l)]$ and the second $[X_2(x/l)]$ modes of free vibrations of the plane cantilever bar (CFRP) with reinforcing angles $\pm 20°$. Nomenclature: E—Eulerian model (without consideration of transverse shear effect), T—Timoshenko model. Bar dimensions: $l = 1$ m, $H = b = 40$ mm (see Figure 9.1).

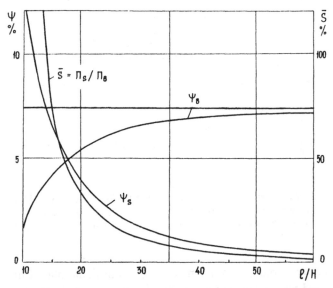

FIGURE 9.5. Ratio between shear energy, Π_S, and bending energy, Π_D ($\bar{S} = \Pi_S/\Pi_D$), dissipation factors owing to shear ($\psi_S = \Delta\Pi_S/\Pi$) and bending ($\psi_D = \Delta\Pi_D/\Pi$) versus relative length of the plane cantilever bar, l/H. The bar is constituted of angle-plied layers with reinforcing angles $\pm 20°$. The straight line corresponds to the dissipation factor value under uniaxial loading. Cross-sectional dimensions: $H = b = 40$ mm (see Figure 9.1).

191

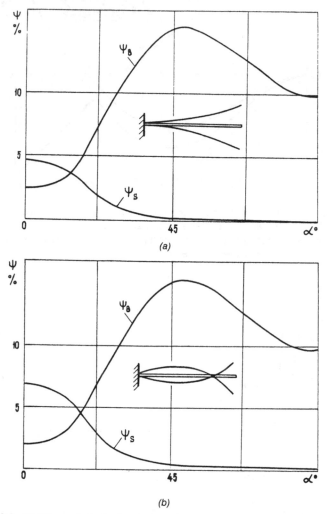

FIGURE 9.6. Variations of dissipation factors due to shear ($\psi_S = \Delta\Pi_S/\Pi$) and bending ($\psi_D = \Delta\Pi_D/\Pi$) with fiber orientation angle, $\pm\alpha$, of plane cantilever bar: (a) the first mode of vibrations and (b) the second form of vibrations. Bar dimensions: $l = 1$ m, $H = b = 40$ mm (see Figure 9.1).

take a unit bar volume that is enclosed by coordinate planes in the cylindrical coordinate system (Figure 9.7). The polar coordinate system for the cross section (r,θ) is also shown there. The stresses that develop due to the differences in the Poisson's ratios of the bar layers can be ignored. In that case the separated element is under uniaxial loading along the x-axis due to the action of normal stresses, σ [Equation (9.27)] and under shear due to

TABLE 9.1. *Predicted Results of Three Vibration Modes for a Plane Cantilever Bar Made of Angle-Plied CFRP with Reinforcing Angles ±20 (Figure 9.1).* *

[. . .]		The 1st Mode	The 2nd Mode	The 3rd Mode
ν_E	Hz	59.1	370.0	1037.0
ν_T	Hz	59.2	354.0	903.0
ψ_D	%	6.21	5.76	3.43
ψ_S	%	2.46	3.42	8.37
ψ	%	8.67	9.18	11.80
N	%/s	513.0	3250.0	10,653
$\bar{S}$	%	18.7	28.0	115.0

*ν_E, ν_T are vibration frequencies in the Eulerian model (without consideration of transverse shear) and in the Timoshenko model respectively; ψ_D is the dissipation factor due to bending; ψ_S is the dissipation factor due to shear; N is the dissipation power coefficient; $\bar{S}$ is the ratio between the shear energy and the bending energy of the bar; $\psi = \psi_D + \psi_S$.
Linear dimensions of the bar: $l = 1$ m, $H = b = 40$ mm, the stiffness $D = 2.5 \times 10^7$ GPa·mm⁴, $k' = 3.6 \times 10^3$ GPa·mm².

the action of shear stresses, τ. Shear stresses are developed in the (x,z) plane due to shear strains, γ [Equation (9.26)].

The integral with respect to normal stresses, σ, determines the bending moment, M, and the integral with respect to shear stresses, τ, determines the transverse force, Q [(Equation 9.3)]. Bar EDC (elasto-dissipative characteristics) owing to the action of the moment, M, and the transverse

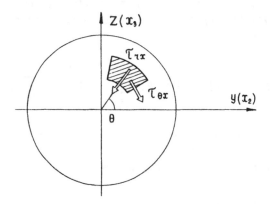

FIGURE 9.7. Tangential stresses in transverse cross section of round bar τ_{rx}, $\tau_{\theta x}$ in cylindrical coordinate system (r,θ,x). Coordinate systems (x,y,z) (Figure 9.2) and (x_1,x_2,x_3) for the bar coincide.

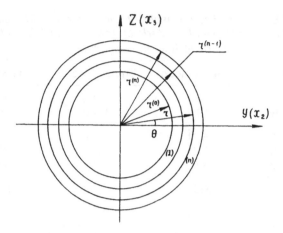

FIGURE 9.8. Boundaries between the layers in the cross section of the round bar.

characteristics) owing to the action of the moment, M, and the transverse force, Q, do not mutually affect each other, i.e., the EDC have the property of associativity.

Let us define the bending stiffness and energy losses caused by bending for the round hollow bar. We will number the layers $(1,. . .,k,. . .,n)$ and the radii of their boundaries, $r^{(k-1)}, r^k$ $(k = 1,2,. . .,n)$ as shown in Figure 9.8. Ignore the stresses that develop due to the difference in the Poisson's ratios of the bar layers. The unit volume under pure bending is then considered to be under a uniaxial stress state along the x-axis, which is the bar axis. The stress, σ, relates to the strain, ϵ, and the bar curvature, $\varkappa$, in the following way: $\sigma = E_1\epsilon = E_1 z\varkappa$, where E_1 is the Young's modulus along the bar axis. The bending moment, M, is defined by the integral over the area S of the bar cross section:

$$M = \int_S \sigma z dS = \left(\sum_{k=1}^{n} E_1^{\{k\}} \int_{S^{(k)}} z^2 dS\right)\varkappa = \sum_{k=1}^{n} d_k\varkappa \qquad (9.40)$$

Here k is the number of the layer (according to Figure 9.8), and the coefficients, d_k, describe bending stiffnesses of each kth layer. The coefficients, d_k, are represented in the polar coordinates as

$$d_k = E_1^{\{k\}} \int_{r^{(k-1)}}^{r^{(k)}} \int_0^{2\pi} r^3 \sin^2 \theta d\theta dr = \frac{E_1^{\{k\}}\pi}{4}[(r^{(k)})^4 - (r^{(k-1)})^4]$$

$$(9.41)$$

The bending stiffness of a round bar is characterized by the sum $D = \sum_{k=1}^{n} d_k$. Energy losses ΔW_D in the piece of the bar with the unit length along the x-axis are determined by the integral

$$\Delta W_D = \int_S \frac{1}{2} \psi_{11} \sigma^2 dS = \frac{1}{2} \varkappa^2 \sum_{i=1}^{n} \psi_{11}^{(k)} (E_1^{(k)})^2 \int_{S^{(k)}} z^2 dS \quad (9.42)$$

Here $\psi_{11}^{(k)}$ is the EDC matrix component of the angle-plied (orthotropic) composite material of the kth layer under a plane stress state [Equation (8.36)]. Young's modulus is related to the component of the layer compliance matrix as follows: $E_1^{(k)} = 1/s_{11}^{(k)}$. Applying Equations (9.40) and (9.41), one will obtain equations for energy losses [Equation (9.42)] and the potential energy, W_D, in the following form

$$\Delta W_D = \frac{1}{2} \varkappa^2 F_D$$

$$(9.43)$$

$$W_D = \frac{1}{2} D \varkappa^2$$

where

$$F_D = \sum_{i=1}^{n} \psi_{11}^{(k)} d_k / s_{11}^{(k)}, \quad D = \sum_{i=1}^{n} d_k$$

Energy losses and the potential energy of the whole bar, considering Equation (9.43), are as follows

$$\Delta \Pi_D = \frac{1}{2} F_D \int_0^l \varkappa^2 dx, \quad \Pi_D = \frac{1}{2} D \int_0^l \varkappa^2 dx \quad (9.44)$$

As it follows from Equation (9.44), the dissipation factor, ψ, does not depend upon the modes of vibrations for the round bar under pure bending

$$\psi = \Delta \Pi_D / \Pi_D = F_D / D \quad (9.45)$$

Shear strains in the cylindrical coordinate system, γ_{rx} and $\gamma_{\theta x}$, are related to the shear strain, γ [Equation (9.26)], as follows

$$\gamma_{rx} = -\gamma \sin \theta, \quad \gamma_{\theta x} = \gamma \cos \theta \quad (9.46)$$

By Hooke's law, shear stresses equal

$$\tau_{rx} = G_{13}\gamma_{rx}, \qquad \tau_{\theta x} = G_{12}\gamma_{\theta x} \qquad (9.47)$$

Here G_{13} and G_{12} are the shear moduli of the bar material for the plane specimen with respect to the coordinate system (x_1, x_2, x_3) that coincides with the system (x, y, z) (see Figure 9.2). Elasto-dissipative characteristics of the unit volume in the cylindrical coordinate system will coincide with corresponding characteristics of the plane bar.

If the layer of the round bar is the angle-plied composite, the modulus G_{13} should be changed according to Equation (9.28). The coefficient, G_{12}, considering Equations (4.103), (8.27), and (8.33), equals

$$G_{12} = (g'_{11} - 2g'_{12} + g'_{22})\,s^2c^2 + (s^2 - c^2)g'_{66} \qquad (9.48)$$

where g'_{ij} are the components of the monolayer stiffness matrix in the natural coordinate system.

The transverse force, Q, is determined by the integral over the area of the cross section, S, with respect to the sum of the projections of the shear stresses

$$Q = \int_S (\tau_{\theta x}\cos\theta + \tau_{rx}\sin\theta)\mathrm{d}S \qquad (9.49)$$

The integration in Equation (9.49) should be performed with respect to the polar coordinate system of the cross section (see Figure 9.7), considering Equations (9.26), (9.46), and (9.47). The expression for the coefficient, k' [Equation (9.3)], is then of the form:

$$k' = \frac{\pi}{2}\sum_{k=1}^{n}[G_{12}^{\{k\}}(r^2 - r^4/8R^2) - G_{13}^{\{k\}}(r^3 - 3r^4/8R^2)]\Big|_{r^{(k-1)}}^{r^{(k)}} \qquad (9.50)$$

Energy losses in the piece of the round bar with the unit length, ΔW_s, owing to shear stresses can be written as

$$\Delta W_s = \frac{1}{2}\int_S (\tau_{rx}^2\psi_{55} + \tau_{\theta x}^2\psi_{66})\mathrm{d}S \qquad (9.51)$$

Here the components of the EDC matrix of the layers are determined from Equations (4.89). For angle-plied layers the coefficients are calculated according to Equations (9.33) and (8.36). Substituting stress values from Equations (9.26), (9.46), and (9.47) in Equation (9.51), it is possible to derive the expression for energy losses as

$$\Delta W_s = \frac{1}{2} F_s \gamma_0^2$$

(9.52)

$$F_s = \sum_{k=1}^{n} (G_{13}^{\{k\}})^2 \psi_{55}^{\{k\}} I_{13}^{\{k\}} + \sum_{k=1}^{n} (G_{12}^{\{k\}})^2 \psi_{66}^{\{k\}} I_{12}^{\{k\}}$$

where the coefficients $I_{13}^{\{k\}}$ and $I_{12}^{\{k\}}$ of the kth layer are the integrals in the polar coordinate system of the bar cross section (see Figure 9.7). These are of the form:

$$I_{13}^{\{k\}} = \int_{r^{(k-1)}}^{r^{(k)}} \int_{0}^{2\pi} \sin^2 \theta \left(1 - \frac{r^2 \sin^2 \theta}{R^2} \right)^2 r\,dr\,d\theta$$

(9.53)

$$I_{12}^{\{k\}} = \int_{r^{(k-1)}}^{r^{(k)}} \int_{0}^{2\pi} \cos^2 \theta \left(1 - \frac{r^2 \sin^2 \theta}{R^2} \right)^2 r\,dr\,d\theta$$

The integrals [Equation (9.53)] are calculated analytically, their values being equal to

$$I_{13}^{\{k\}} = \frac{\pi r^2}{2} \left(1 - \frac{3r^2}{4R^2} + \frac{5r^4}{24R^4} \right) \Bigg|_{r^{(k-1)}}^{r^{(k)}}$$

(9.54)

$$I_{12}^{\{k\}} = \frac{\pi r^2}{2} \left(1 - \frac{r^2}{4R^2} + \frac{r^4}{24R^4} \right) \Bigg|_{r^{(k-1)}}^{r^{(k)}}$$

Complete energy losses, $\Delta \Pi$, in the round cantilever bar are calculated from Equation (9.35), taking into account Equations (9.43) and (9.52).

The bar potential energy, Π, equals the sum of the shear potential energy, Π_S, and the bending potential energy, Π_D. To calculate the value of Π, it is sufficient to apply the formulas obtained for $\Delta \Pi$. In doing so, some transformations must be done. The components ψ_{55} and ψ_{66} should be changed to $1/G_{13}$ and $1/G_{12}$ respectively in Equations (9.35) and (9.52). The part of the potential energy caused by the action of the bending moment, Π_D, is calculated from Equation (9.44).

In Equation (9.35) the integral with respect to the square of the bar curvature is calculated analytically by the following formula

$$\int_0^l x^2 dx = A^2 \frac{m^3}{2} [2ml\beta^2 + (\beta^2 - 1) \cos ml (\sin ml + 2 \sinh ml)$$

$$+ (\beta^2 + 1) \times \cosh ml(2 \sin ml + \text{sh } ml) + \beta(\text{ch } 2ml - \cos 2ml)$$

$$+ 4\beta \sin ml \sinh ml] \qquad (9.55)$$

where to define the values of m and β, one should apply Equations (9.24) and (9.25).

The value of the integral with respect to the square of the shear angle, γ_0, is determined in every specific case by approximate integration, considering that

$$\gamma_0 = A[m_2(\cos m_2 x - c_2 \sin m_2 x) + m_1(c_3 \cosh m_1 x - c_2 \sinh m_1 x)$$

$$+ m\{\cosh mx - \cos mx + \beta(\sin mx + \sinh mx)\}] \qquad (9.56)$$

Let us note that the dissipation factor of the system depends upon the modes and the frequencies of vibrations when transverse shear deformation is considered.

CHAPTER 10

Problems of Rational Reinforcement

Control of dissipation parameters is a rather important and practical subject of inquiry for engineers. Examples can be found where a designer tries both to increase and to decrease dissipation parameters. The problems involved in searching the range of composites or composite structures for those which provide the desired dissipative behavior will be called the problems of rational design for energy dissipation.

10.1 RATIONAL DESIGN OF MULTILAYERED MATERIALS

Let us consider the problem of selecting the multilayered composite structure which will achieve the desired dissipation properties. We will examine two possible formulations of the problem and the related rational criteria (the criteria of rationality).

10.1.1 Formulation 1

Suppose that a designer must determine the interior structure of a material which will give the minimum amplitude for a set number of vibration cycles.

Let us first use the energy balance Equation (9.1). The dissipation factor was identified as the ratio of the energy losses in a vibration cycle to the value of the energy amplitude in the cycle

$$\psi = \Delta\Pi/\Pi \qquad (10.1)$$

Using this definition, one can obtain the following equation:

$$(A')^2 = A^2 - \psi(A + A')^2/4 \tag{10.2}$$

Here A and A' are the initial and the final vibration amplitudes in the cycle respectively. Further transformation of Equation (10.2) results in the expression

$$A' = \frac{4 - \psi}{4 + \psi} A \tag{10.3}$$

Equation (10.3) shows that the sequence of the amplitudes forms a geometric progression with a denominator that equals $(4 - \psi)/(4 + \psi)$.

Analysis of Equation (10.3) indicates that the requirement that the amplitude be a minimum for a given number of vibration cycles, n, calls for the maximum value of the dissipation factor, ψ:

$$\min A(n) \rightarrow \max \psi \tag{10.4}$$

10.1.2 Formulation 2

Suppose that it is now necessary to provide the minimum amplitude of vibrations in a definite time.

Let us present the envelope for the sequence of amplitudes (see Figure 2.3) in the form:

$$A = A_0 e^{-Nt/2} \tag{10.5}$$

The parameter, N, will be determined from the equality between the denominator of the geometrical progression [Equation (10.3)] and the decrement of vibrations, $e^{-Nt/2}$,

$$N = -\frac{2}{T} \ln \frac{4 - \psi}{4 + \psi} \tag{10.6}$$

where T is the period of vibrations

The requirement that the amplitude should be a minimum in a definite time [Equation (10.5)] reduces to the requirement that the parameter, N [Equation (10.6)], reach a maximum:

$$\min A(t) \rightarrow \max N \tag{10.7}$$

Let us exchange Equation (10.6) for the approximate equation. Take into account the first term of the expansion in series for the logarithm in Equation (10.6), and exchange the period, *T,* for the vibration frequency, ω, according to the formula

$$T = 2\pi/\omega$$

Then one obtains

$$N = \psi\omega/2\pi \tag{10.8}$$

Multiplying the dissipation factor by the vibration frequency defines the dissipation power. That is why the physical meaning of Equation (10.7) is the minimum of the dissipated power.

Now let us express the vibration frequency, ω, in terms of the stiffness of the angle-plied bars, using Equations (9.22) and (9.36):

$$\omega = m^2\sqrt{D/\varrho} = L/\sqrt{s_{11}} \tag{10.9}$$

L is a parameter that does not depend on the layers EDC, and the reinforcing angle, s_{11}, is the component of the compliance matrix for the angle-plied composite [see Equation (8.39)]. The component s_{11} is the inverse of the corresponding modulus of elasticity.

Thus, Equation (10.7) reduces to the maximum value of the product of the dissipation factor multiplied by the square root of the elastic modulus:

$$\max N \rightarrow \max \psi\sqrt{E} \tag{10.10}$$

The criteria of Equations (10.4) and (10.10) can be applied to produce approximate designs of angle-plied materials and thin-walled bars by direct use of the diagrams from Chapters 4 and 8.

For example, the dissipation factor, ψ, and Young's modulus, *E,* are plotted in Figure 8.7 as the functions of the orientation angle, α, for the angle-plied CFRP HTS/DX 210 under uniaxial loading. The values of the dissipation factor peak at $\alpha \approx 55°$. This value of the orientation angle is the optimal one according to the criterion of Equation (10.4). Using the criterion of Equation (10.10), one must take into account the dependencies of the dissipation factor, ψ, and Young's modulus, *E,* upon the angle, α, at one time. The value of the rational orientation angle decreases as compared to the first case and is in the vicinity of the angle $\alpha \approx 30°$. More exact determination of optimal structures can be realized with the help of standard methods of mathematical programming.

10.2 DESIGN OF A TWO-LINKED MANIPULATOR WITH CONTROLLED DISSIPATIVE RESPONSE

As an example of a more complicated technical system, let us now consider a two-linked manipulator. The manipulator consists of two hollow (tubular) bars. One bar is rigidly fastened together with a load and the opposite end of another bar is rigidly clamped [Figure 10.1(a)]. The point where the load is fastened to the bar coincides with the load's center of

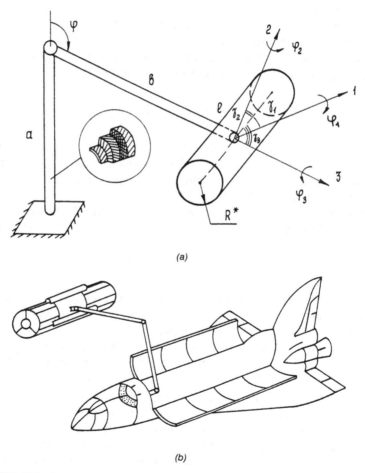

(a)

(b)

FIGURE 10.1. General view of the two-linked manipulator (a), the vehicle of "Space Shuttle" type with the two-linked manipulator (b).

mass. The angle between the bars is designated as φ. The angle, φ, and the load orientation with respect to the bars can vary during manipulator operation.

Manipulator bars are thin-walled multilayered cylinders which consist of n orthotropic composite layers with a cylindrical orthotropy of their properties [Figure 10.1(a)]. The bar that is closest to the fixing point has a length a. The bar that is fastened to the load has a length b. The load is supposed to be a solid uniform round cylinder of radius R, length l, and mass m.

The system represented is a simplified model of the manipulator used in the vehicles of "Space Shuttle" (U.S.A.) and "Buran" (USSR) type [Figure 10.1(b)].

The dissipative behavior of the system described is responsible, in many cases, for its quality. The time taken to execute the operations in orbit (and hence the cost of the project) depends upon the dissipative behavior. The accuracy of manipulator operation (for example, the minimum amplitude of controlled load displacements) is closely connected with the dissipative response of the system.

It is reasonable to assume that the dissipative behavior of the system is not the only factor that must be taken into account when designing such complicated systems. Proper attention should be paid to the system stiffness characteristics, the frequencies of the vibrations, and the strengths of its elements at all stages of the vehicle flight.

10.2.1 Modes and Frequencies of Natural Vibrations of the System

Let us consider free damped vibrations of the system for fixed parameters of the load orientation and the bar geometry. Assume that the weight characteristics of the bars are several orders less than the load weight, so that the bars are believed to be weightless. A perfectly rigid load has six degrees of freedom which correspond to progressive displacements along coordinate axes and rotation around the axes.

The orientation of the Cartesian coordinate system is chosen as follows: its origin is placed in the load's center of mass; axis 1 is directed perpendicularly to the plane of the manipulator bars; axis 3 is directed along the bar of length b; and axis 2 is in the plane of manipulator bars, perpendicular to axes 1 and 3 as shown in Figure 10.1(a). The load is a cylinder, which may be oriented arbitrarily with respect to the coordinate axes. The load position is given by the angles γ_1, γ_2, and γ_3, which are the angles between the load longitudinal axis and coordinate axes 1, 2, and 3. Evidently the values of the angles are related in the following way: $\cos^2 \gamma_1 + \cos^2 \gamma_2 + \cos^2 \gamma_3 = 1$.

The equation for free vibrations of the system is represented in the form [82]

$$V_i + \beta_{ij} m_j \ddot{V}_j = 0 \tag{10.11}$$

Here $\ddot{V}_j = d^2 V_j/dt^2$, V_i are the linear and angular load displacements with respect to coordinate axes. Let us assume that V_1, V_2, and V_3 correspond to linear displacements of the load along coordinate axes (1,2,3), and V_4, V_5, V_6 are the angles of rotation ($\varphi_1, \varphi_2, \varphi_3$) with respect to these axes respectively.

Induction coefficients, β_{ij}, relate load factors, P_j (the forces and the moments applied to the load), to the displacements, so that

$$V_i = \beta_{ij} P_j \tag{10.12}$$

The values of β_{ij} are determined with the help of More's integrals in accordance with a well-known scheme [82]. The relations inverse to the relations of Equation (10.12) are of the form

$$P_j = C_{ij} V_i$$

where the parameters C_{ij} are called the stiffness coefficients.

The first three coefficients in Equation (10.11) — m_1, m_2, and m_3 — represent the load mass, m, and the following three coefficients — m_4, m_5, and m_6 — are the inertia moments of the load (I_1, I_2, I_3) with respect to coordinate axes, so that

$$m_1 = m, \qquad m_2 = m, \qquad m_3 = m$$
$$m_4 = I_1, \qquad m_5 = I_2, \qquad m_6 = I_3 \tag{10.13}$$

The solution of Equations (10.11) is represented as follows

$$V_i = a_i \sin \omega t \quad (i = 1,2,\ldots,6) \tag{10.14}$$

Once Equation (10.14) is substituted into Equation (10.11), one will obtain a set of uniform linear equations with respect to the amplitudes, a_i, of system free vibrations

$$a_i - \omega^2 \beta_{ij} m_j a_j = 0 \quad (i = 1,2,\ldots,6) \tag{10.15}$$

Let us represent Equation (10.15) in the matrix form

$$\{a\} - \omega^2[\beta][E]\{m\}\{a\}^\mathsf{T} = 0 \qquad (10.16)$$

where, considering Equation (10.13), the column matrices of displacement amplitudes $\{a\}$ and the mass-column matrices $\{m\}$ are of the forms

$$\{a\} = \begin{Bmatrix} a_1 \\ a_2 \\ a_3 \\ a_4 \\ a_5 \\ a_6 \end{Bmatrix}, \qquad \{m\} = \begin{Bmatrix} m_1 \\ m_2 \\ m_3 \\ m_4 \\ m_5 \\ m_6 \end{Bmatrix} = \begin{Bmatrix} m \\ m \\ m \\ I_1 \\ I_2 \\ I_3 \end{Bmatrix}$$

$[E]$ is the unit matrix of the size (6×6), $[\beta]$ is the symmetrical matrix (6×6) of the induction coefficients. To obtain the values of the stiffness coefficients, C_{ij}, it is sufficient to determine the inverse matrix $[\beta]^{-1}$ which will coincide with the corresponding stiffness matrix.

Based on the rules of transformation for the inertia tensor of the solids, when rotating the coordinate system, the following relations will be obtained

$$I_1 = (I_z - I_x)\cos^2\gamma_1 + I_x = 0$$

$$I_2 = (I_z - I_x)\cos^2\gamma_2 + I_x = 0 \qquad (10.17)$$

$$I_3 = (I_z - I_x)\cos^2\gamma_3 + I_x = 0$$

where

$$I_x = I_y = \frac{m}{4}(R_*^2 + l^2/3)$$

$$I_z = mR_*^2/2$$

I_x, I_y, and I_z in Equation (10.17) are the inertial moments of the load-cylinder with respect to main central inertial axes (the axis, z, is the longitudinal cylinder axis).

The condition for the existence of a nontrivial solution of combined uniform Equations (10.15) is that the system determinant equals zero. This condition is the secular equation with respect to the frequencies of system vibrations

$$\begin{vmatrix} -\dfrac{1}{\omega^2} + \beta_{11}m_1 & \beta_{12}m_2 & \beta_{13}m_3 & \beta_{14}m_4 & \beta_{15}m_5 & \beta_{16}m_6 \\[2mm] \beta_{12}m_1 & -\dfrac{1}{\omega^2} + \beta_{22}m_2 & \beta_{23}m_3 & \beta_{24}m_4 & \beta_{25}m_5 & \beta_{26}m_6 \\[2mm] \beta_{13}m_1 & \beta_{23}m_2 & -\dfrac{1}{\omega^2} + \beta_{33}m_3 & \beta_{34}m_4 & \beta_{35}m_5 & \beta_{36}m_6 \\[2mm] \beta_{14}m_1 & \beta_{24}m_2 & \beta_{34}m_3 & -\dfrac{1}{\omega^2} + \beta_{44}m_4 & \beta_{45}m_5 & \beta_{46}m_6 \\[2mm] \beta_{15}m_1 & \beta_{25}m_2 & \beta_{35}m_3 & \beta_{45}m_4 & -\dfrac{1}{\omega^2} + \beta_{55}m_5 & \beta_{56}m_6 \\[2mm] \beta_{16}m_1 & \beta_{26}m_2 & \beta_{36}m_3 & \beta_{46}m_4 & \beta_{56}m_5 & -\dfrac{1}{\omega^2} + \beta_{66}m_6 \end{vmatrix} = 0$$

$$(10.18)$$

The matrix that is under the sign of the system determinant [Equation (10.18)] is not symmetrical, as the coefficients m_i are different. Secular Equation (10.18) is a sixth degree equation with respect to ω^2. It has six roots, each of which defines the natural frequency of the system. The solution of combined Equations (10.15) corresponds to the value of the natural vibration frequency, ω_i. This solution is the matrix of vibration amplitudes $\{a_i\}$. This matrix is known as the modes of vibrations. The modes of vibrations, because of the linear dependence of Equations (10.15), are determined with accuracy to a constant factor and are mutually orthogonal.

Let us derive the expressions for the components of the induction coefficients matrix for the case of the two-linked manipulator with a load that has six degrees of freedom. It is well known that the induction coefficient, β_{ij}, is the displacement along the direction of the ith loading factor under the action of the unit jth factor. Let us turn our attention to More's integrals. Preliminary estimates for real systems showed that the effect of longitudinal and transverse forces on the modes and the frequencies of vibrations and on the energy dissipation might be ignored in thin-walled round tubular bars with considerable length. Thus More's integrals are of the form

$$\beta_{ij} = \int_a \frac{M_{T_i}M_{T_j}}{T^a}dz + \int_b \frac{M_{T_i}M_{T_j}}{T^b}dz + \int_a \frac{M_{x_i}M_{x_j}}{D^a}dz$$

$$(10.19)$$

$$+ \int_b \frac{M_{x_i}M_{x_j}}{D^b}dz + \int_a \frac{M_{y_i}M_{y_j}}{D^a}dz + \int_b \frac{M_{y_i}M_{y_j}}{D^b}dz$$

Here Cartesian coordinate axes (x, y, z) (in the cross sections of the bars) are selected so that the z-axis is the longitudinal one, the x-axis is perpendicular to the plane of bars system, and the y-axis is perpendicular to the x- and z-axes. The integration is performed over the lengths of the bars a and b. The nomenclature in Equation (10.19) is: T^a and T^b are the torsional stiffness characteristics of the bars; D^a and D^b are the bending stiffnesses with respect to the x- or y-axes; M_{T_i} is the torsional moment; M_{x_i} and M_{y_i} are the bending moments with respect to the x- or y-axes in the bar cross sections. The moments M_{x_i} and M_{y_i} arise from the action of the ith unit force factor applied in the load's center of mass.

The values of the integrals in Equation (10.19) are calculated by the standard procedure, multiplying bending-moment diagrams by torsional-moment diagrams. Corresponding diagrams are illustrated by Figure 10.2.

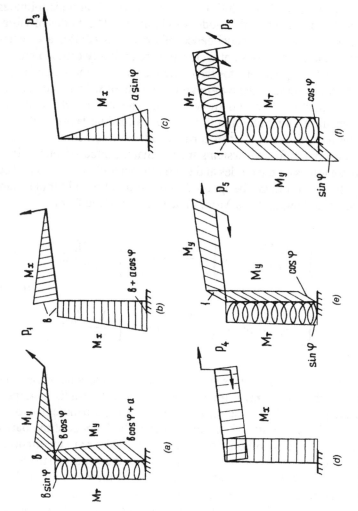

FIGURE 10.2. Diagrams of bending moment (M_{x_i}, M_{y_i}) and torsion moment (M_{T_i}) caused by the action of unit force factors.

The following expressions will be derived for the induction factors

$$\beta_{11} = \frac{a^3}{3D^a} + \frac{b^3}{3D^b} + ab^2\left(\frac{\cos^2\varphi}{D^a} + \frac{\sin^2\varphi}{T^a}\right) + \frac{b\cos\varphi a^2}{D^a}$$

$$\beta_{12} = 0, \qquad \beta_{13} = 0, \qquad \beta_{14} = 0$$

$$\beta_{15} = \frac{a^2\cos\varphi}{2D^a} + \frac{b}{2D^b} + ab\left(\frac{\cos^2\varphi}{D^a} + \frac{\sin^2\varphi}{T^a}\right)$$

$$\beta_{16} = \frac{ab\sin 2\varphi}{2T^a} - \frac{ab\sin 2\varphi + a^2\sin\varphi}{2D^a}$$

$$\beta_{22} = \frac{a^3\cos^2\varphi}{3D^a} + \frac{b^3}{3D^b} + \frac{a^2b\cos\varphi + b^2a}{D^a}$$

$$\beta_{23} = -\frac{3a^2b\sin\varphi + a^3\sin 2\varphi}{6D^a}$$

$$\beta_{24} = \frac{ab - (1/2)a^2\cos\varphi}{D^a} - \frac{b^2}{2D^b}$$

$$\beta_{25} = 0, \qquad \beta_{26} = 0$$

$$\beta_{33} = \frac{a^3\sin^2\varphi}{3D^a}, \qquad \beta_{34} = \frac{a^2\sin\varphi}{2D^a}$$

$$\beta_{35} = 0, \qquad \beta_{36} = 0$$

$$\beta_{44} = \frac{a}{D^a} + \frac{b}{D^b}, \qquad \beta_{45} = 0, \qquad \beta_{46} = 0$$

$$\beta_{55} = \frac{a\cos^2\varphi}{D^a} + \frac{b}{D^b} + \frac{a\sin^2\varphi}{T^a}$$

$$\beta_{56} = \frac{a\sin 2\varphi}{2}\left(\frac{1}{T^a} - \frac{1}{D^a}\right)$$

$$\beta_{66} = \frac{a\cos^2\varphi}{T^a} + \frac{b}{T^b} + \frac{a\sin^2\varphi}{D^a}$$

(10.20)

Considering Equation (10.20), the induction coefficients matrix takes the form

$$[\beta] = \begin{bmatrix} \beta_{11} & 0 & 0 & 0 & \beta_{15} & \beta_{16} \\ 0 & \beta_{22} & \beta_{23} & \beta_{24} & 0 & 0 \\ 0 & \beta_{23} & \beta_{33} & \beta_{34} & 0 & 0 \\ 0 & \beta_{24} & \beta_{34} & \beta_{44} & 0 & 0 \\ \beta_{15} & 0 & 0 & 0 & \beta_{55} & \beta_{56} \\ \beta_{16} & 0 & 0 & 0 & \beta_{56} & \beta_{66} \end{bmatrix} \tag{10.21}$$

The matrix $[\beta]$ [Equation (10.21)] includes two groups of the components. One of them corresponds to induction coefficients when the manipulator is loaded by two forces along axes 2 and 3 (in the bars plane) and the moment with respect to axis 1. The second one corresponds to the "anti-plane" loading. Then, the determinant of the matrix [Equation (10.18)] is represented as the product of two third-order determinants, taking into account substitution of the coefficients, m_i, in accordance with Equation (10.13), i.e.,

$$\det\,(p[E] + [\alpha^1]) \cdot \det\,(p[E] + [\alpha^2]) = 0 \tag{10.22}$$

where $p = -1/\omega^2$, the matrices

$$[\alpha^1] = \begin{bmatrix} \beta_{11}m & \beta_{15}I_2 & \beta_{16}I_3 \\ \beta_{15}m & \beta_{55}I_2 & \beta_{56}I_3 \\ \beta_{16}m & \beta_{56}I_2 & \beta_{66}I_3 \end{bmatrix}$$

$$[\alpha^2] = \begin{bmatrix} \beta_{22}m & \beta_{23}m & \beta_{24}I_1 \\ \beta_{23}m & \beta_{33}m & \beta_{34}I_1 \\ \beta_{24}m & \beta_{34}m & \beta_{44}I_1 \end{bmatrix} \tag{10.23}$$

The secular Equation (10.22) is equivalent to two independent cubic equations with respect to the parameter, p. These equations are of the form

$$p^3 + rp^2 + sp + t = 0 \tag{10.24}$$

Here the coefficients of the equation are expressed in terms of the components of the matrices $[\alpha^1]$ and $[\alpha^2]$ as follows

$$r = -(\alpha^i_{11} + \alpha^i_{22} + \alpha^i_{33})$$

$$s = \alpha^i_{22}\alpha^i_{33} + \alpha^i_{11}(\alpha^i_{22} + \alpha^i_{33}) - \alpha^i_{13}\alpha^i_{31} - \alpha^i_{23}\alpha^i_{32} - \alpha^i_{12}\alpha^i_{21}$$

$$\tag{10.25}$$

$$t = \alpha^i_{11}\alpha^i_{23}\alpha^i_{32} + \alpha^i_{22}\alpha^i_{13}\alpha^i_{31} + \alpha^i_{33}\alpha^i_{12}\alpha^i_{21} - \alpha^i_{11}\alpha^i_{22}\alpha^i_{33}$$

$$- \alpha^i_{12}\alpha^i_{23}\alpha^i_{31} - \alpha^i_{13}\alpha^i_{21}\alpha^i_{32} \quad (i = 1,2)$$

The cubic Equation (10.24) can be transformed by changing the variables $p = y - (r/3)$. It can then be represented in the following simpler form

$$y^3 + \tau y + q = 0 \qquad (10.26)$$

where $\tau = (3s - r^3)/3$, $q = 2r^3/27 - rs/3 + t$.

In the case of three real roots, the standard solution of Equation (10.26) is written in the form

$$y_1 = -2\varrho \cos \theta/3$$

$$y_2 = -2\varrho \cos (\theta/3 + 2\pi/3) \qquad (10.27)$$

$$y_3 = -2\varrho \cos (\theta/3 + 4\pi/3)$$

Here $\varrho = (\sin q)\sqrt{|\tau|/3}$, $\cos \theta = q/2\varrho^3$.

Just as secular Equation (10.22) is divided into two independent cubic equations, so the set of Equations (10.16) in vibration modes has two groups of independent solutions of the form:

$$\{a\} = \{a_1,0,0,0,a_5,a_6\}^{\mathrm{T}}, \qquad \{a\} = \{0,a_2,a_3,a_4,0,0\}^{\mathrm{T}}$$

Since all vibration amplitudes are proportional to the common parameter, then the value of any amplitude is assumed to be equal to unity, for example

$$\{a\} = \{1,0,0,0,a_5,a_6\}^{\mathrm{T}}$$

$$\qquad (10.28)$$

$$\{a\} = \{0,1,a_3,a_4,0,0\}^{\mathrm{T}}$$

The other two values of a_i for both groups of vibration modes are determined from the system of nonuniform second-order equations

$$\gamma_{15}a_5 + \gamma_{16}a_6 = -\gamma_{11}$$

$$\qquad (10.29)$$

$$\gamma_{55}a_5 + \gamma_{56}a_6 = -\gamma_{51}$$

$$\gamma_{23}a_3 + \gamma_{24}a_4 = -\gamma_{23}$$

$$\qquad (10.30)$$

$$\gamma_{33}a_3 + \gamma_{34}a_4 = -\gamma_{32}$$

where γ_{ij} are the components of the matrix $[\gamma]$ that is expressed in terms of the frequency parameter $p = -1/\omega^2$ and the matrices $[\alpha^1]$, $[\alpha^2]$

$$[\gamma] = \begin{bmatrix} p[E] + [\alpha^1] & 0 \\ 0 & p[E] + [\alpha^2] \end{bmatrix} \qquad (10.31)$$

Solutions of the sets of Equations (10.29) and (10.30) are of the form:

$$a_5 = \frac{-\gamma_{11} - \gamma_{16}a_6}{\gamma_{15}}, \qquad a_6 = \frac{\gamma_{55}\gamma_{11} - \gamma_{15}\gamma_{51}}{\gamma_{56}\gamma_{15} - \gamma_{55}\gamma_{16}}$$

$$a_3 = \frac{-\gamma_{22} - \gamma_{24}a_4}{\gamma_{23}}, \qquad a_4 = \frac{\gamma_{33}\gamma_{22} - \gamma_{23}\gamma_{32}}{\gamma_{34}\gamma_{23} - \gamma_{33}\gamma_{24}} \qquad (10.32)$$

To define energy losses and the level of the system potential energy in a loading cycle, it will be necessary to know the vibration modes in terms of force and moment amplitudes. Based on Equations (10.12), (10.21), and (10.28), one will obtain

$$\{P\} = [\beta]^{-1}\{a\} \qquad (10.33)$$

Here, when inverting the two-block matrix $[\beta]$, it is sufficient to determine inverse matrices for two third-order submatrices. Column matrices of the forces and the moments, $\{P\}$, which correspond to Equation (10.28), take the following form

$$\{P\} = \{P_1,0,0,0,P_5,P_6\}^T, \qquad \{P\} = \{0,P_2,P_3,P_4,0,0\}^T \qquad (10.34)$$

Without disturbing the generality of the manipulations, all the modes of vibrations (in terms of the forces and the moments) may be divided by one of the components, for example

$$\{1,0,0,0,F_5,F_6\} = \{1,0,0,0,P_5/P_1,P_6/P_1\}$$

$$\{0,1,F_3,F_4,0,0\} = \{0,1,P_3/P_1,P_4/P_1,0,0\} \qquad (10.35)$$

10.2.2 Energy Dissipation in the System

Let us define the parameters of the relative energy dissipation for a vibrating two-linked manipulator, i.e., the ratios of system energy losses in a loading cycle to amplitude values of the total potential energy. The calculation is performed for six modes of natural vibrations. There is a need to derive the expressions for energy losses and the potential energy of the system, if torsion and bending of round tubular multilayered bars are taken into account. Energy losses and the potential energy caused by shear stresses (torsion) and normal stresses (pure bending) can be separated. As for consideration of transverse shear, the estimates showed that the corrections are small in the case of composite bars with real geometrical dimensions and reinforced structures (such estimates were carried out by the procedure described in Chapter 9, Section 9.3). The corrections came to only 1–2% for the energy losses and still less for the frequency and the stiffness. That is why they will not be taken into account.

Let us consider an element that is cut from the bar layer by two meridian and two equatorial sections, as is shown in Figure 10.3. In accordance with the assumptions described above, the element is under uniaxial loading along the bar axis (normal stresses, σ, caused by the bending moment) and under shear in the layer plane (shear stresses, τ, caused by the torsional moment). Additional stresses arising due to the difference between the Poisson's ratios of the bar layers are ignored. Energy losses in a unit volume of the bar, ΔW, are represented as

$$\Delta W = \Delta W^* + \Delta W^{**}$$

$$\Delta W^* = \frac{1}{2} \psi_{11}^{(k)} \sigma^2, \qquad \Delta W^{**} = \frac{1}{2} \psi_{66}^{(k)} \tau^2$$

(10.36)

Here $\psi_{11}^{(k)}$ and $\psi_{66}^{(k)}$ are the components of the EDC stress matrix of the kth layer in the bar coordinate system.

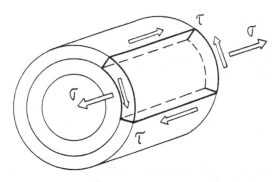

FIGURE 10.3. Stress state of unit volume separated by coordinate surfaces in a tubular specimen.

Let us determine energy losses in a piece of the bar of unit length. The losses are divided into the energy lost owing to bending, ΔW_D, and the energy lost owing to torsion, ΔW_T. The first term, ΔW_D, is known, since it was determined in Chapter 9, Section 9.3 [see Equation (9.43)]. Let us now consider the energy dissipation owing to torsion. We will change to shear strains in Equation (10.36) on the basis of Hooke's law: $\tau = g_{66}^{(k)} \gamma$, where $g_{66}^{(k)}$ is the component of the stiffness matrix of the kth layer in the global coordinate system. The losses, ΔW_T, are defined by the integral of specific energy losses, ΔW^{**}, over the area, S, of the bar cross section

$$\Delta W_T = \int_S \Delta W^{**} dS \qquad (10.37)$$

The angle of shear, γ, depends linearly upon the radius, ϱ, and it is related to the angle of twist, θ, by the relationship

$$\gamma = \varrho\theta \qquad (10.38)$$

Then, upon integrating over the bar layers, Equation (10.37) takes the form

$$\Delta W_T = \frac{\theta^2}{2} F_T \qquad (10.39)$$

where

$$F_T = \sum_{k=1}^{n} \psi_{66}^{(k)} T_k / s_{66}^{(k)}$$

Here T_k is the torsional stiffness, $s_{66}^{(k)}$, is the component of the compliance matrix of the kth layer

$$T_k = \frac{1}{2} \pi g_{66}^{(k)} [(r^{(k)})^4 - (r^{(k-1)})^4], \qquad s_{66}^{(k)} = 1/g_{66}^{(k)} \qquad (10.40)$$

To calculate the potential energy, it is sufficient to change the component of the EDC matrix, $\psi_{66}^{(k)}$, to the component of the compliance matrix, $s_{66}^{(k)}$, in Equations (10.39) and (10.40). Finally the amplitude value of the torsion potential energy will be equal to

$$W_T = \frac{\theta^2}{2} T, \qquad T = \sum_{k=1}^{n} T_k \qquad (10.41)$$

where T is the torsional stiffness of the bar.

Let us write down the relationships between the angle of twist, θ, and the torsional moment, M_T, between the curvature, x, and the bending moment, M:

$$\theta = M_T/T, \qquad x = M/D \qquad (10.42)$$

Here D is the bar bending stiffness [Equation (9.43)]. Energy losses $\Delta\Pi$ and the potential energy, Π, of the manipulator link are defined by the integrals [Equations (9.44), (10.39), and (10.41)] over the length, L:

$$\Delta\Pi = \frac{1}{2}\int_{L} F_D x^2 dz + \frac{1}{2}\int_{L} F_T \theta^2 dz$$

$$\Pi = \frac{1}{2}\int_{L} D x^2 dz + \frac{1}{2}\int_{L} T\theta^2 dz$$

(10.43)

Considering the fact that stiffness characteristics and dissipation parameters do not vary within the length of each bar (a and b), Equation (10.43) takes the form

$$\Delta\Pi = \frac{1}{2} \sum_{i=a,b} (F_D^i I_D^i + F_T^i I_T^i)$$

$$\Pi = \frac{1}{2} \sum_{i=a,b} (D^i I_D^i + T^i I_T^i)$$

(10.44)

where F_D^i, F_T^i, D^i, and T^i ($i = a,b$) are dissipation and stiffness parameters [Equations (9.43), (10.39), and (10.40)] for the bars a and b; the other

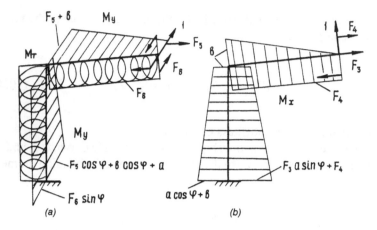

FIGURE 10.4. Diagrams of bending moment (M_{x_i}, M_{y_i}) and torsion moment (M_{T_i}) caused by the action of unit force factors. Diagrams correspond to two vibration modes of the manipulator: (a) $(1,0,0,0,F_5,F_6)$ and (b) $(0,1,F_3,F_4,0,0)$.

coefficients depend upon the modes of vibrations and are defined in terms of the bending moment, M_D, and the torsion moment, M_T, according to Equation (10.42)

$$I_D^i = \int_i \frac{M^2}{D^2}\,dz, \qquad I_T^i = \int_i \frac{M_T^2}{T^2}\,dz, \quad (i = a, b) \qquad (10.45)$$

When calculating the values of the parameters I_D and I_T [Equation (10.45)], let us use the modes of vibrations in the forces [Equation (10.35)]. The corresponding bending-moment diagrams and torsional-moment diagrams are shown in Figure 10.4. Note that the bending of each bar takes place only in one coordinate plane. When integrating in accordance with Equations (10.45), one defines the parameters which correspond to the vibration mode $(1,0,0,0,F_5,F_6)$:

$$I_D^a = \frac{c^2a + ca^2 + a^3/3}{(D^a)^2}, \qquad I_D^b = \frac{bF_5^2 + F_5b^2 + b^3/3}{(D^b)^2}$$

$$c = (b + F_5)\cos\varphi - F_6 \sin\varphi \qquad (10.46)$$

$$I_T^a = \frac{(b \sin\varphi + F_5 \sin\varphi + F_6 \cos\varphi)^2 a}{(T^a)^2}, \qquad I_T^b = \frac{F_6^2 b}{(T^b)^2}$$

and to the mode $(0,1,F_3,F_4,0,0)$:

$$I_D^a = \frac{a^3 A^2/3 + a^2 AB + B^2 a}{(D^a)^2}$$

$$A = F_3 \sin \varphi - \cos \varphi, \qquad B = F_4 - b$$

$$I_D^b = \frac{b^3/3 - b^2 F_4 + F_4^2 b}{(D^b)^2}$$

$$I_T^a = 0, \qquad I_T^b = 0$$

(10.47)

The energy dissipation in the system is characterized by the relative dissipation, $\psi = \Delta\Pi/\Pi$, and the power factor of dissipation, N, which depends on the frequency of vibrations, ω: $N = \psi\omega/2\pi$. The first parameter defines the energy dissipation in a complete cycle of vibrations, the second one describes the vibration damping in a unit of time.

10.2.3 Criterion of Rationality

Definition of the quality criterion (i.e., the answer to the question: "What is good and what is bad?") is a nontrivial problem for any complex technical system. In the case of the space manipulator, this definition relates not only to the internal parameters of the manipulator itself, but to the parameters of the control system as well, and even to the tactics of orbit operations. Nonetheless, let us try to describe the main features of the problem.

Let us consider that the manipulator mass was reserved in the design of the whole space vehicle, and the problem with manipulator design is how best to use its mass limit. Execution time for orbit operations is the dominant parameter of the system. What this means is that primary attention should be paid to the power factor of the system energy dissipation, N, which defines the vibration damping in a unit of time

$$N = \psi\omega/2\pi$$

(10.48)

It follows directly from Equation (10.48) that a rise in N can be obtained both due to direct increase in the dissipation parameter, ψ, and due to increase in the frequency (or the stiffness) of the system.

Orbit operations can be performed at different mutual positions of the manipulator bars. The angle between the bars, φ, defines their position. Six frequencies of load natural vibrations, ω_i, and six values of the dissipation factor, ψ_i, correspond to every φ value. The following problem formulations are possible:

(1) Maximize the power factor of the energy dissipation in the limited narrow range for the angles, φ. This is the local optimization. In this case, the astronauts should try to carry out the most exacting operations at a definite mutual arrangement of manipulator bars. It is clear that such a technique restricts significantly the abilities of the space vehicle as a whole.

(2) Maximize the average value of the power factor of the energy dissipation for all positions of the bars and all vibration modes. This technique is a rational one, which provides reasonable average damping level. However, some mutual arrangements of the bars can have intolerably low values of the damping parameters.

(3) Maximize the minimum value (among all possible bars' positions and vibration modes) of the power factor. This variant of the optimization criterion provides some assured minimum of damping parameters for the astronauts and control system designers.

Let us stop at the last criterion. The problem is providing:

$$\max \, (\min N(\varphi,\omega_i)) \quad i = 1,2,\ldots,6$$

when the internal structure of the bars (i.e., the thicknesses and orientation angles of the monolayers) is variable

$$\overline{h}^{(k)}, \, \pm\alpha^{(k)} \rightarrow \mathrm{var} \quad i = 1,2,\ldots,n \tag{10.49}$$

where n is the number of different layers in the bars.

Bending and torsion stiffnesses are constrained

$$D \geq D_0, \qquad T \geq T_0$$

The geometrical parameters of the bars are assumed to be given. It is also assumed that the other constraints (e.g., for the bars' strength) are met.

10.2.4 Example of Design

Let us consider the problem of the rational reinforcing of two-linked manipulator bars. The objective is to obtain accelerated damping of free vibrations with respect to all modes for all allowable values of the angle, φ, between the links [see Figure 10.1(a)]. The angle, φ, varies through the range from $0°$ to $160°$. The load is a ball of 3.1 m diameter, its mass equals unity. Manipulator links are similar cylindrical hollow bars 6 m in length, with an internal radius of 150 mm. They consist of seven angle-plied layers 0.3 mm in thickness each. The monolayers are made of high modulus carbon fiber reinforced plastic "Kulon," CFRP properties are given in Chapter 3, Table 3.1.

Selection of the rational bar structure is performed by examining all possible versions according to the following algorithm. A data file with the information on reinforced structures is set up. Each structure is described by five parameters. These are the four numbers of the layers (with reinforcing angles given as subscripts)

$$n_0, n_{90}, n_{45}, n_\alpha \qquad (10.50)$$

and the reinforcing angle in the "controlling" layer, $\pm\alpha$. A number of the layers of each type, n_i, takes the values from zero to seven. The total number of layers in the bar wall is equal to seven, i.e.,

$$n_0 + n_{90} + n_{45} + n_\alpha = 7 \qquad (10.51)$$

At the first design stage the values of the reinforcing angle, $\pm\alpha$, in the "controlling" layer are set up with a step equal to $10°$ varying in the range

$$0° \leq |\alpha| \leq 90° \qquad (10.52)$$

The bending stiffness, D, and torsion stiffness, T, are then calculated for every version from the data file for reinforced structures. If these stiffness characteristics do not meet the requirements

$$D \geq D_0, \qquad T \geq T_0 \qquad (10.53)$$

the version under examination is considered to be unsatisfactory and is not analyzed further. Here D_0 and T_0 are the bending and torsion stiffness characteristics of some "basic" version for the bar's structure. The following version is considered to be the "basic" one in the present example:

$$n_0 = 3, \qquad n_{90} = 1, \qquad n_{45} = 1, \qquad n_\alpha = 2, \qquad \alpha = \pm 25°$$

$$(10.54)$$

The total thickness of the bar wall (2.1 mm) is two orders less than the bar diameter (~ 300 mm). One can then ignore the individual position of each layer in the bar wall, when determining the stiffness characteristics and dissipation parameters.

If the reinforcing structure meets the requirements of Equation (10.53), then one should define the frequencies, the modes of vibrations, the relative energy dissipation, and the power coefficient of energy dissipation for nine discrete values of the angle between the manipulator links, φ, with the step $\Delta\varphi = 10°$.

Calculation results are estimated by the criterion of Equation (10.49). Further calculations are performed for the rational reinforced structure (defined at the first design stage) with a fixed number of layers [Equation (10.50)]. The value of $\pm\alpha$ varies with the step $\Delta\alpha = \pm 1°$. The best [from the standpoint of the criterion of Equation (10.49)] reinforced structure of the second stage is believed to be the final design.

Calculation results for 10 types of reinforced structures are tabulated in Table 10.1. The first version is the "basic" one, the second version is the rational (recommended) one, and the third version is the worst with respect to the damping behavior. Here ψ_D and ψ_T are the dissipation factors of the bars under bending and torsion. They are determined by the ratio of energy losses (under corresponding loading conditions) in the bar volume of the unit length to the amplitude value of the potential energy. The values min $N(\varphi)$ and max $N(\varphi)$ are the minimum and maximum values of the power coefficient of dissipation when changing the angle, φ, between manipulator links.

The rational structure consists of similar angle-plied layers with reinforcing angles $\alpha = \pm 19°$. Compared to the "basic" version, the stiffness characteristics of the rational bar (especially torsional stiffness) increased, the value of the energy dissipation factor under bending increased more than 30%, and the value of the energy dissipation factor under torsion was lowered somewhat. Note that though the difference in minimum values of the power coefficient of dissipation, N, for the "basic" and the rational versions is minor, the maximum values of N differ essentially (by a factor of 1.5). This is an advantage of the rational reinforced structure. Figures 10.5–10.10 illustrate the plots of the relative energy dissipation of the two-linked manipulator, ψ, the power coefficient of dissipation, N, the cyclic frequency ν ($\nu = \omega/2\pi$) versus the angle between the manipulator links, φ, for the "basic" and the rational structures. Figures 10.5–10.7 correspond to three vibration modes of the type $(1,0,0,0,a_5,a_6)$, while the other Figures 10.8–10.10 correspond to the modes $(0,1,a_3,a_4,0,0)$. For the last modes of vibrations the dissipation factor, ψ, does not depend on the angle, φ, as the energy dissipates in the bars only due to bending. Unfavorable values of the

TABLE 10.1. *Elasto-Dissipative Characteristics of Two-Linked Manipulator for Different Versions of Bar Reinforced Structures.*

Parameters	Version Number									
	1	2	3	4	5	6	7	8	9	10
n_0	3	0	4	2	1	4	3	3	2	0
n_{90}	1	0	0	1	1	0	0	0	0	0
n_{45}	1	0	1	2	2	2	3	1	2	2
n_α	2	7	2	2	3	1	1	3	3	5
$\pm\alpha°$	25	19	60	10	10	10	20	30	10	10
$D \cdot 10^8$, GPa·mm⁴	28	29	31	30	29	38	28	28	37	35
$T \cdot 10^8$, GPa·mm⁴	9.7	11.4	10.9	9.9	10.3	9.5	14.2	13.7	10.3	11.0
ψ_D, %	4.9	6.89	3.91	4.15	4.26	3.94	4.67	5.12	4.10	4.27
ψ_T, %	5.13	4.79	4.88	5.08	5.00	5.18	4.41	4.48	5.00	4.84
min $N(\varphi)$, %/S	48	51	43	44	44	45	46	47	45	46
max $N(\varphi)$, %/S	1290	1842	1094	1137	1151	1217	1233	1345	1241	1265

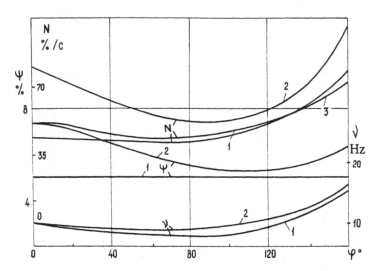

FIGURE 10.5. Dissipation factor, ψ, power coefficient of dissipation, N, frequencies, ν, versus an angle, φ, between manipulator links. 1—Basic version of bar reinforced structures, 2—rational version, 3—version with the worst dissipation. The mode of vibrations is of $(1,0,0,0, a_s^{(1)}, a_6^{(1)})$ type.

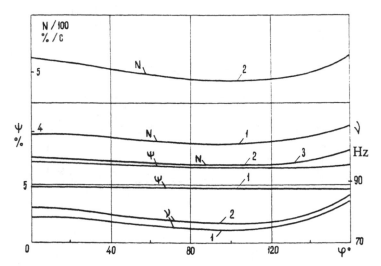

FIGURE 10.6. Dissipation factor, ψ, power coefficient of dissipation, N, frequencies, ν, versus an angle, φ, between manipulator links. 1—Basic version of bar reinforced structures, 2—rational version, 3—version with the worst dissipation. The mode of vibrations is of $(1,0,0,0, a_s^{(2)}, a_6^{(2)})$ type.

222

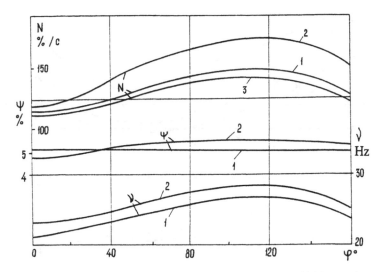

FIGURE 10.7. Dissipation factor, ψ, power coefficient of dissipation, N, frequencies, ν, versus an angle, φ, between manipulator links. 1—Basic version of bar reinforced structures, 2—rational version, 3—version with the worst dissipation. The mode of vibrations is of $(1,0,0,0,a_5^{(3)},a_6^{(3)})$ type.

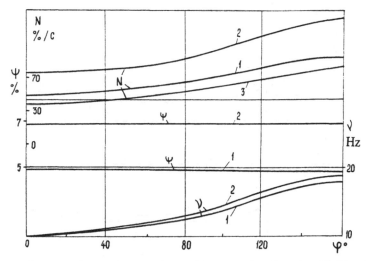

FIGURE 10.8. Dissipation factor, ψ, power coefficient of dissipation, N, frequencies, ν, versus an angle, φ, between manipulator links. 1—Basic version of bar reinforced structures, 2—rational version, 3—version with the worst dissipation. The mode of vibrations is of $(0,1,a_3^{(4)},a_4^{(4)},0,0)$ type.

223

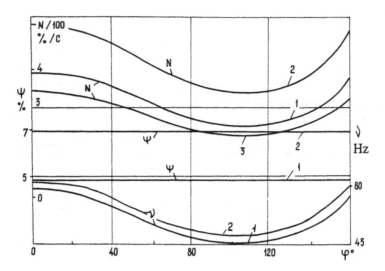

FIGURE 10.9. Dissipation factor, ψ, power coefficient of dissipation, N, frequencies, ν, versus an angle, φ, between manipulator links. 1—Basic version of bar reinforced structures, 2—rational version, 3—version with the worst dissipation. The mode of vibrations is of $(0,1,a_3^{(5)},a_4^{(5)},0,0)$ type.

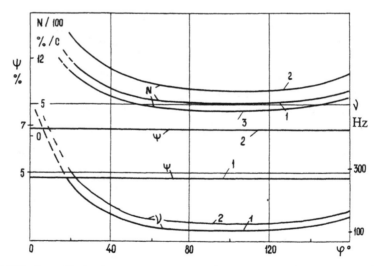

FIGURE 10.10. Dissipation factor, ψ, power coefficient of dissipation, N, frequencies, ν, versus an angle, φ, between manipulator links. 1—Basic version of bar reinforced structures, 2—rational version, 3—version with the worst dissipation. The mode of vibrations is of $(0,1,a_3^{(6)},a_4^{(6)},0,0)$ type.

224

power coefficient of dissipation correspond to the first vibration mode in the vicinity of the angle $\varphi \approx 90°$, as can be seen from Figure 10.10.

The plots in Figures 10.5–10.10 indicate that the rational reinforced structure has a significant advantage in damping behavior over the basic version. The plots of ψ and N for the rational structure are placed far above the analogous plots for the basic version.

Tensors in the Rectangular Cartesian Coordinate System

Tensors are special mathematical objects that have properties independent of the coordinate systems used for their description. In a local coordinate system, a tensor is defined as a set of quantities (numbers) called its components. Whether any set of these numbers defines the tensor depends on a rule of the numbers transformation when changing from one coordinate system to another.

Let the coordinate axes, x_1, x_2, x_3, have the unit direction vectors (or the basis) $\bar{e}_1, \bar{e}_2, \bar{e}_3$.

The tensor of rank k is the set of quantities depending on k indices $A_{i_1 i_2, \ldots i_k}$ so that under the coordinate system transformation (i.e., when changing from the basis $(\bar{e}_1, \bar{e}_2, \bar{e}_3)$ to the basis $\bar{e}_1', \bar{e}_2', \bar{e}_3'$, and vice versa) tensor components in the new coordinate system $A_{i_1 i_2, \ldots i_k}'$ are related to the components in the old system by the following formulas

$$A_{i_1 i_2 \ldots i_k}' = \alpha_{i_1 j_1} \alpha_{i_2 j_2} \ldots \alpha_{i_k j_k} A_{j_1 j_2 \ldots j_k}$$

$$A_{i_1 i_2 \ldots i_k} = \alpha_{j_1 i_1} \alpha_{j_2 i_2} \ldots \alpha_{j_k i_k} A_{j_1 j_2 \ldots j_k}$$

(A.1)

Here α_{ij} are the direction cosines of the angles between the vectors $\bar{e}_i$ and $\bar{e}_j$, i.e.,

$$\alpha_{ij} = (\bar{e}_i \cdot \bar{e}_j)$$

(A.2)

In Equation (A.1) and further the convention is applied to summation with respect to repeated indices varying from 1 to 3. For example, $a_i b_i$ is interpreted as follows

$$a_i b_i = a_1 b_1 + a_2 b_2 + a_3 b_3$$

(A.3)

The vector $\bar{a} = a_i \bar{e}_i$ is a particular example of a tensor of rank 1. Actually vector components in an arbitrary new basis $(\bar{e}_1', \bar{e}_2', \bar{e}_3')$ and in the old one $(\bar{e}_1, \bar{e}_2, \bar{e}_3)$ are related by the formulas, analogous to Equation (A.1),

$$a_i' = \alpha_{ij} a_j \quad \text{and} \quad \alpha_i = \alpha_{ji} a_j' \tag{A.4}$$

Tensor $A_{i_1 i_2, \ldots i_k}$ of rank k and tensor $B_{j_1 j_2, \ldots j_p}$ of rank p can form a new multiplicative tensor of rank $k + p$ according to the following rule

$$C_{i_1 i_2 \ldots i_k j_1 j_2 \ldots j_p} = A_{i_1 i_2 \ldots i_k} B_{j_1 j_2 \ldots j_p} \tag{A.5}$$

Equation (A.5) can be interpreted as a multiplication of the tensors. For example, if there is a pair of vectors, a_i and b_j, then the quantities $a_i b_j$ are the components of the multiplicative tensor of the second rank, i.e.

$$c_{ij} = a_i b_j \tag{A.6}$$

Contraction of the tensor over two free indices is the operation whereby two indices are designated by one and the same letter, and therefore they become the summation indices. As a result of contraction the tensor is obtained. The order of this tensor is two units less than the order of the initial one. So, the summation over two repeated indices, when contracting the tensor of order two $c_{ij} = a_i b_j$, results in a scalar quantity

$$c_{ii} = a_i b_i = a_1 b_1 + a_2 b_2 + a_3 b_3 \tag{A.7}$$

The scalar does not depend on coordinate system selection, i.e., it is the invariant of the coordinate system transformation. Carrying out the operations of tensor contraction and forming the multiplicative tensor, it is possible to obtain the scalars, i.e., invariant quantities. Thus, multiplying the tensor σ_{ij} by itself and contracting it, the following invariants can be obtained.

$$S_1 = \sigma_{ii}$$

$$S_2 = \sigma_{ij}\sigma_{ij} \tag{A.8}$$

$$S_3 = \sigma_{ij}\sigma_{ik}\sigma_{jk}$$

The criterion for tensor character of the quantities is formulated as follows: a set of quantities $A_{i_1 i_2 \ldots i_k}$ forms a tensor of rank k, if their contraction (according to the rules for the tensors) with the tensor $B_{j_1 j_2 \ldots j_p}$ of rank p results in a tensor of rank $k - p$.

For example, if the quantities a_i and b_j are vectors (tensors of order 1), then the set of quantities c_{ij} is a tensor of second rank, provided the contraction $c_{ij} a_i b_j$ is a scalar quantity.

This criterion is called the indirect test for tensor character.

The tensor is called symmetric with respect to a pair of indices if it does not vary when the indices are transposed. The tensor is called antisymmetric if it changes its sign under the transposition of a pair of the indices. For example, the tensor A_{ijkl} is symmetric with respect to the pair of the indices ij, if

$$A_{ijkl} = A_{jikl} \tag{A.9}$$

The tensor of second rank B_{ij} is antisymmetric, if

$$B_{ij} = -B_{ji} \tag{A.10}$$

Any second rank tensor can be resolved into symmetric and antisymmetric parts

$$A_{ij} = \frac{1}{2}(A_{ij} + A_{ji}) + \frac{1}{2}(A_{ij} - A_{ji}) \tag{A.11}$$

Matrices

B.1 FUNDAMENTAL DEFINITIONS

Matrix $[A]$ is a system of elements (numbers, for example) arranged in a certain order and forming a rectangular table which consists of m rows and n columns.
Matrix $[A]$ is written as follows

$$[A] = \begin{bmatrix} a_{11} & a_{12} & \cdots & a_{1n} \\ a_{21} & a_{22} & \cdots & a_{2n} \\ \cdots & \cdots & \cdots & \cdots \\ a_{m1} & a_{m2} & \cdots & a_{mn} \end{bmatrix} \tag{B.1}$$

There is a conception for matrix size. Matrix $[A]$ for example is of size m by n.
If the number of matrix columns n equals 1, the matrix is called a column matrix, and it looks like

$$[A] = \begin{Bmatrix} a_1 \\ a_2 \\ \cdot \\ \cdot \\ \cdot \\ a_m \end{Bmatrix} \tag{B.2}$$

If all elements of a matrix equal zero, it is called a zero matrix, and it is written as

$$[A] = [0] \tag{B.3}$$

If the number of matrix rows equals the number of matrix columns, the matrix is called a square one. For a square matrix the term "matrix order" is used along with the term "matrix size." So a matrix of size m by m has the mth order.

230

It is possible to distinguish the main diagonal in a square matrix. It is formed by the elements with equal indices: $a_{11}, a_{22}, \ldots, a_{mm}$. A square matrix, whose elements all equal zero except for the main diagonal elements, is called a diagonal matrix

$$[A] = \begin{bmatrix} a_{11} & 0 & \cdots & 0 \\ 0 & a_{22} & \cdots & 0 \\ \cdots & \cdots & \cdots & \cdots \\ 0 & 0 & \cdots & a_{mm} \end{bmatrix} \tag{B.4}$$

If all nonzero elements of the diagonal matrix are equal one to another:

$$a_{11} = a_{22} = \ldots = a_{mm} = a$$

then the matrix is called a scalar one. If nonzero elements are equal to unity in the scalar matrix, the matrix is called the unit (identity) one and is designated by

$$[E] = \begin{bmatrix} 1 & 0 & \cdots & 0 \\ 0 & 1 & \cdots & 0 \\ \cdots & \cdots & \cdots & \cdots \\ 0 & 0 & \cdots & 1 \end{bmatrix} \tag{B.5}$$

Any number can be considered a square matrix of order 1. Such a matrix is called a scalar.

A square matrix is called symmetric if its elements arranged symmetrically about the main diagonal are equal one to another

$$\alpha_{ij} = \alpha_{ji} \tag{B.6}$$

A matrix $[A]$ can be partitioned by horizontal and/or vertical lines into several smaller rectangular parts, which are matrices too. So, the matrix

$$[A] = \begin{bmatrix} a_{11} & a_{12} & a_{13} & \vdots & a_{14} & \vdots & a_{15} & a_{16} \\ a_{21} & a_{22} & a_{23} & \vdots & a_{24} & \vdots & a_{25} & a_{26} \\ \cdots & \cdots & \cdots & \cdot & \cdots & \cdot & \cdots & \cdots \\ a_{31} & a_{32} & a_{33} & \vdots & a_{34} & \vdots & a_{35} & a_{36} \end{bmatrix}$$

is divided into the blocks $[A_{11}], [A_{12}], [A_{13}], [A_{21}], [A_{22}], [A_{23}]$

$$[A_{11}] = \begin{bmatrix} a_{11} & a_{12} & a_{13} \\ a_{21} & a_{22} & a_{23} \end{bmatrix} \quad [A_{12}] = \begin{bmatrix} a_{14} \\ a_{24} \end{bmatrix} \quad [A_{13}] = \begin{bmatrix} a_{15} & a_{16} \\ a_{25} & a_{26} \end{bmatrix}$$

$$[A_{21}] = [a_{31} a_{32} a_{33}] \quad [A_{22}] = [a_{34}] \quad [A_{23}] = [a_{35} a_{36}]$$

Now matrix $[A]$ can be represented as a matrix whose elements are also the matrices

$$[A] = \begin{bmatrix} A_{11} & A_{12} & A_{13} \\ A_{21} & A_{22} & A_{23} \end{bmatrix} \tag{B.7}$$

In this case matrix $[A]$ is called the block matrix and the matrices $[A_{11}]$, $[A_{12}]$, . . .,$[A_{23}]$ are called the submatrices or the blocks.

The determinant of a square matrix is a number calculated from matrix elements according to the following rules:

- The determinant of order n equals the algebraic sum of n − factorial members.
- Each term of the sum is the product of n matrix elements singly taken as the factors from each row and each column of the table.
- The sum term has a plus sign, if the permutations formed by the first and the second indices of its elements, a_{ij}, have the same evenness (both are even or both are odd), otherwise it has a minus sign.
- The permutation of n numbers $1,2,. . .,n$ is any of their arrangements; two numbers in the permutation form an inversion (a deviation from the order), if the larger number is placed ahead of the smaller one; the permutation is called even, if a number of its inversions is even.

According to this definition the determinant of order 3 equals

$$\det A = \begin{vmatrix} a_{11} & a_{12} & a_{13} \\ a_{21} & a_{22} & a_{23} \\ a_{31} & a_{32} & a_{33} \end{vmatrix}$$

$$= a_{11}a_{22}a_{33} + a_{12}a_{23}a_{31} + a_{13}a_{21}a_{32} - a_{13}a_{22}a_{31} - a_{11}a_{23}a_{32} - a_{12}a_{21}a_{33} \tag{B.8}$$

The square matrix determinant is designated by the symbol

$$\begin{vmatrix} a_{11} & a_{12} & . . . & a_{1n} \\ a_{21} & a_{22} & . . . & a_{2n} \\ . . . & . . . & . . . & . . . \\ a_{n1} & a_{n2} & . . . & a_{nn} \end{vmatrix}$$

or briefly as $|A|$, $\det A$, ΔA.

It is possible to delete some rows and some columns in the rectangular matrix $[A]$ so that one obtains a square matrix of order k with undeleted elements, which is called a matrix minor of order k.

The number r, equal to the greatest order of the nonzero determinant produced by the matrix $[A]$, is called the rank of the matrix $[A]$.

If the matrix $[A]$ is a square one, then the minors, diagonal elements of which are the diagonal elements of the matrix $[A]$, are called the main minors.

A square symmetric matrix whose main minor determinants all are positive is called a positively defined matrix. A positively defined matrix is in correspondence with the positively defined quadratic form

$$f = \sum_{i=1}^{n} \sum_{j=1}^{n} a_{ij} x_{ij} \tag{B.9}$$

The quadratic form has positive values at all real (not equal to zero simultaneously) values of x_i and x_j. Similarly, the definition is introduced for the nonnegatively defined square symmetric matrix and the corresponding quadratic form.

B.2 OPERATIONS ON MATRICES

Two matrices are equal if they are of the same size and all their corresponding elements are equal in pairs, i.e.

$$[A] = [B]$$

$$\text{if } a_{ij} = b_{ij} \quad (i = 1,\ldots,m; j = 1,\ldots,n) \tag{B.10}$$

The sum of two matrices of equal orders $[A]$ and $[B]$ is the matrix of the same order, whose elements are defined by the equations

$$c_{ij} = a_{ij} + b_{ij} \tag{B.11}$$

The next rules follow from the matrix sum definition

$$[A] + [B] = [B] + [A]$$

$$([A] + [B]) + [C] = [A] + ([B] + [C]) \tag{B.12}$$

$$[A] + [0] = [A]$$

Multiplication of matrix $[A]$ by matrix $[B]$ is reasonable, only if the number of columns in matrix $[A]$ is equal to the number of rows in matrix $[B]$.

If matrix $[A]$ is of the size $(m \times p)$ and matrix $[B]$ is of the size $(p \times n)$, then the product of matrix $[A]$ multiplied by matrix $[B]$ is matrix $[C]$ of the size $(m \times n)$

$$[C] = [A][B]$$

The elements of the matrix $[C]$ are defined by the equation

$$c_{ij} = a_{i1}b_{1j} + a_{i2}b_{2j} + \ldots + a_{ip}b_{pj} = \sum_{k=1}^{p} a_{ik}b_{kj}$$

(B.13)

$$(i = 1,\ldots,m; j = 1,\ldots,n)$$

In the general case the product of the matrices is not commutative, i.e.

$$[A][B] \neq [B][A] \qquad \text{(B.14)}$$

The main properties of the matrices' product (if the matrix sizes satisfy the definition of multiplication operation) are the following

$$[A][0] = [0][A] = [0]$$

$$[A][E] = [E][A] = [A]$$

$$([A] + [B])[C] = [A][C] + [B][C]$$
$$[A]([B] + [C]) = [A][B] + [A][C]$$

(B.15)

$$([A][B])[C] = [A]([B][C])$$

$$\det ([A][B]) = \det [A] \det [B]$$

The product of multiplying matrix $[A]$ by the scalar a is matrix $[B]$, each element of which is equal to the product of multiplying the corresponding element of the matrix $[A]$ by the scalar α

$$b_{ij} = \alpha a_{ij} \qquad \text{(B.16)}$$

The next rules are valid

$$(\alpha + \beta)[A] = \alpha[A] + \beta[A]$$

$$\alpha([A] + [B]) = \alpha[A] + \alpha[B]$$

(B.17)

$$(\alpha\beta)[A] = \alpha(\beta[A]) = \beta(\alpha[A])$$

$$\alpha([A][B]) = (\alpha[A])[B] = [A](\alpha[B])$$

The matrix transposition is the operation of replacing the matrix's rows by its columns, provided their numbers are kept constant. The matrix thus obtained from the matrix $[A]$ is called the transposed matrix with respect to the matrix $[A]$ and is designated by $[A]^{\mathrm{T}}$.

The matrix $\{A\}^T$ obtained by the transposition of the column matrix $\{A\}$ is called the row matrix

$$\{A\}^T = \{a_1, a_2, \ldots, a_i\}$$

The transposition of the block matrix is performed as follows:

$$[A]^T = \begin{bmatrix} a_{11} & a_{12} \\ a_{21} & a_{22} \\ a_{31} & a_{32} \end{bmatrix}^T = \begin{bmatrix} a_{11}^T & a_{21}^T & a_{31}^T \\ a_{12}^T & a_{22}^T & a_{32}^T \end{bmatrix}$$

The transposition of the symmetric matrix gives the initial one.
The transpose operation has the following properties

$$([A] + [B])^T = [A]^T + [B]^T$$

$$(\alpha[A])^T = \alpha[A]^T$$

$$([A][B])^T = [B]^T[A]^T \qquad\qquad (B.18)$$

$$([A]^T)^T = [A]$$

If $[A]$ is a square matrix and matrix $[B]$ is such that

$$[A][B] = [B][A] = [E]$$

then $[B]$ is called an inverse matrix with respect to matrix $[A]$, and it is designated as $[A]^{-1}$. By definition

$$[A][A]^{-1} = [A]^{-1}[A] = [E] \qquad\qquad (B.19)$$

Matrix $[A]$ has an inverse matrix $[A]^{-1}$ if and only if it is nonsingular, i.e., $\det [A] \neq 0$.
The following equations are valid

$$([A]^{-1})^{-1} = [A]$$

$$([A][B])^{-1} = [B]^{-1}[A]^{-1}$$

$$([A]^{-1})^T = ([A]^T)^{-1} \qquad\qquad (B.20)$$

$$\det [A]^{-1} = 1/\det [A]$$

Fourier Transforms

The Fourier transforms are closely related to the Fourier integrals. The function $F(y)$ that corresponds to the function $f(x)$ according to the formula

$$F(y) = \frac{1}{\sqrt{2\pi}} \int_{-\infty}^{+\infty} f(x)e^{iyx}dx \tag{C.1}$$

is called the transformant function. The transfer from the function $f(x)$ to the integral function $F(y)$ is called the Fourier transform.

If Fourier transformant function $F(y)$ is specified, then, through the so-called inverse Fourier transform,

$$f(x) = \frac{1}{\sqrt{2\pi}} \int_{-\infty}^{+\infty} F(y)e^{-ixy}dy \tag{C.2}$$

one will obtain the function $f(x)$ again. In general, $F(y)$ is the complex function for all real $f(x)$. The initial function $f(x)$ can take complex values, having the real argument x.

C.1 THEOREM OF CONVOLUTION

Fourier transformant function of two functions convolution

$$(f * g)(x) = \int_{-\infty}^{+\infty} f(y)g(x - y)dy$$

236

is equal, with an accuracy of $\sqrt{2\pi}$, to the product of the Fourier transformant functions of the multipliers

$$\frac{1}{\sqrt{2\pi}} \int_{-\infty}^{+\infty} (f * g)(x)e^{ixy}dx = \sqrt{2\pi}F(y)G(y) \qquad (C.3)$$

C.2 THEOREM OF DERIVATION

On a range of Fourier transform values, the derivation operation transforms into multiplication by an independent variable, i.e., if $G(y)$ is the Fourier transformant function of $f'(x)$, then $G(y) = -iyF(y)$, where $F(y)$ is the Fourier transformant function of $f(y)$.

Second Order Curved Lines

In the plane of real axes (x,y), a second order curve is the set of points with the coordinates which satisfy the equation:

$$ax^2 + 2bxy + cy^2 + 2dx + 2ey + f = 0$$

where

$$a^2 + b^2 + c^2 \neq 0$$

Once a curve is brought to the canonical form, the curves can be classified in the following way.

(1) *Central curves* (there is a center of symmetry). The general equation of the curve in the canonical form is

$$\lambda_1 x^2 + \lambda_2 y^2 + g = 0$$

where $\lambda_1 > 0$.
Classification follows Table D.1.

TABLE D.1. Classification for Central Curves.

λ_2	g	Curve Form
>0	<0	Ellipse
>0	>0	The equation does not have a solution in real numbers
>0	$=0$	Single point $(0,0)$
<0	$\neq 0$	Hyperbola
<0	$=0$	Pair of intersecting straight lines

TABLE D.2. Classification for Parabolic Curves.

h	k	Curve Form
$\neq 0$	any	Parabola
$= 0$	< 0	Two straight lines parallel to the y-axis
$= 0$	$= 0$	Double straight line (the y-axis)
$= 0$	> 0	No solution

(2) *Parabolic lines* (there is no center of symmetry). The general equation of the curve in the canonical form is

$$\lambda_1 x^2 + 2hy + k = 0$$

where $\lambda_1 > 0$.

Classification follows Table D.2.

REFERENCES

1. Adams, R. D. and D. G. C. Bacon. 1973. "The Dynamic Properties of Unidirectional Fibre Reinforced Composites in 'Flexure and Torsion,'" *J. Composite Materials*, 7(1):53–67.
2. Adams, R. D. and D. G. C. Bacon. 1973. "Effect of Fibre Orientation and Laminate Geometry on the Dynamic Properties of CFRP," *J. Composite Materials*, 7(5):402–428.
3. Adams, R. D. and D. G. C. Bacon. 1973. "Measurement of Flexural Damping Capacity and Dynamic Young's Modulus of Metals and Reinforced Plastics," *J. Phys. D., Appl. Phys.*, 6:26–41.
4. Adams, R. D., M. A. O. Fox, R. J. L. Fload, et al. 1969. "The Dynamic Properties of Unidirectional Carbon and Glass Fiber Reinforced Plastics in Torsion and Flexure," *J. Composite Materials*, 3(5):594–603.
5. Alam, N. and N. T. Asnani. 1984. "Vibration and Damping Analysis of a Multilayered Cylindrical Shell. Part 1: Theoretical Analysis," *AIAA Journal*, 22(6):803–810.
6. Alam, N. and N. T. Asnani. 1984. "Vibration and Damping Analysis of a Multilayered Cylindrical Shell. Part 2: Numerical Results," *AIAA Journal*, 22(7):975–981.
7. Alfutov, N. A., P. A. Zinoviev and B. G. Popov. 1984. *Analysis of Multilayered Composite Plates and Shells* (in Russian). Moscow: Mashinostroenie Publishing House.
8. Ambartsumyan, S. A., ed. 1967. *Theory of Anisotropic Plates* (in Russian). Moscow: Nauka.
9. Ashkenazi, E. K. and E. V. Ganov. 1980. *Anisotropy of Structural Materials* (Handbook) (in Russian). Leningrad: Mashinostroenie Publishing House.
10. Ashley, H. 1984. "On Passive Damping Mechanisms in Large Space Structures," *Journal of Spacecraft and Rockets*, 21(5):448–455.
11. Baker, W. E., W. E. Woolam and D. Young. 1967. "Air and Internal Damping of Thin Cantilever Beams," *Int. J. Mechanical Science*, 9(11):743–766.
12. Barker, A. I. and H. Vangerko. 1983. "Temperature Dependent Dynamic Shear Properties of CFRP," *Composites*, 14:141–144.
13. Bert, C. W. 1980. *Damping Applications for Vibration Control, ASME AMD-38.* ASME, pp. 53–63.

241

14. Bert, C. W. 1980. *Recent Adv. Struct. Dyn. Pap. Int. Conf.*, Southampton, pp. 693–712.

15. Biderman, V. L., ed. 1980. *Theory of Mechanical Vibrations* (in Russian). Moscow: Vysshaya Shkola.

16. Bolotin, V. V. and Yu. N. Novitchkov. 1980. *Mechanics of Multilayered Structures* (in Russian). Moscow: Mashinostroenie Publishing House.

17. Cardon, A. and C. Hiel. 1981. *Composite Structures, Proceedings of the 1st Int. Conf. Paislay*, September 16–18, 1981, London, New Jersey, pp. 301–311.

18. Chou, T-W., ed. 1992. *Microstructural Design of Fiber Composites*. Cambridge: Cambridge University Press.

19. Christensen, R. M. 1979. *Mechanics of Composite Materials*. New York: John Wiley & Sons, Inc.

20. 1987. *Composite Calculations (COMPCAL), Users Manual*. Lancaster, PA: Technomic Publishing Co., Inc.

21. Dudek, T. J. 1970. "Determination of the Complex Modulus of Viscoelastic Two-Layer Composite Beams," *J. Composite Materials*, 4(1):74–89.

22. Dudek, T. J. 1970. "Young's and Shear Moduli of Unidirectional Composites by a Resonant Beam Method," *J. Composite Materials*, 4(2):232–241.

23. Ernest, B. and J. Daxon. 1975. "Real and Imaginary Parts of the Complex Viscoelastic Modulus for Boron Fiber Reinforced Plastics (BFRP)," *J. Acoust. Soc. Am.*, 57(4): 891–898.

24. Filin, A. P. 1966. *Matrices in Statics of Bar Systems* (in Russian). Leningrad, Moscow: Publishing House for the Literature on Building (Industry).

25. Georgi, H. 1979. "Damping Effects in Aerospace Structures," *Proceedings of 48th Meet. AGARD*, Williamsburg, USA, April 2–3, 1979, pp. 9.1–9.20.

26. Gibson, R. F. and R. Plunkett. 1977. "A Forced-Vibration Technique for Measurement of Material Damping," *Experimental Mechanics* (4):297–302.

27. Gibson, R. F. and R. Plunkett. 1976. "Dynamic Mechanical Behavior of Fiber Reinforced Composites," *J. Composite Materials*, 10(5):325–341.

28. Gibson, R. F. and A. Yau. "Complex Moduli of Chopped Fiber and Continuous Fiber Composites: Comparison of Measurements with Estimated Bounds," *J. Composite Materials*, 14(2):155–167.

29. Gibson, R. F. et al. 1987. *Proceedings of Adv. Mater. Technol. '87: 32nd Int. SAMPE Symp. and Exhib.* Anaheim, CA, Covina, CA, April 6–9, 1987, pp. 231–244.

30. Goldenblat, I. I., ed. 1970. *Plates and Shells of Glass Fiber Reinforced Plastics* (in Russian). Moscow: Mashinostroenie Publishing House.

31. Hashin, Z. 1970. "Complex Moduli of Viscoelastic Composites. I. General Theory and Application to Particulate Composites," *Int. J. Solids Structures*, 6(5):539–552.

32. Hashin, Z. 1970. "Complex Moduli of Viscoelastic Composites. II. Fiber Reinforced Materials," *Int. J. Solids Structures*, 6(6):797–807.

33. Hil'chevskii, V. V. and V. G. Dubenets. *Energy Dissipation in Cyclic Deformation of Materials under Complex Stress State* (in Russian). Kiev: Visha Shkola.

34. Hoa, S. V. and D. Ouellete. 1984. "Damping of Composite Materials," *Polymer Composites*, 5:334–338.

35. Hwang, S. J. and R. F. Gibson. 1987. "Micromechanical Modeling of Damping in Discontinuous Fiber Composites Using a Strain Energy/Finite Element Approach," *J. Engineering Materials and Technology*, 109:47–52.

36. Jones, R. M. 1975. *Mechanics of Composite Materials*. New York: Scripta Book Co.

37. Kochneva, L. F., ed. 1979. *Internal Friction in Solids under Vibrations* (in Russian). Moscow: Nauka.

38. Kruklin'sh, A. A. and A. E. Paeglitis. 1988. "Relative Energy Damping in Laminated Reinforced Plastics," *Mekhanika Kompozitnykh Materialov* (in Russian) (3):449–456.

39. Lee, C. Y., D. S. Thompson and M. V. Gandhi. 1988. "Temperature-Dependent Dynamic Mechanical Properties of Polymeric Laminated Beams," *Trans. ASME: J. Eng. Mater. and Technol.*, 110(2):174–179.

40. Lekhnitskii, S. G., ed. 1963. *Theory of Elasticity of an Anistropic Elastic Body*. San Francisco: Holden-Day.

41. Liaa, D. X., C. K. Sung and B. S. Thompson. 1986. "The Optimal Design of Symmetric Laminated Beams Considering Damping," *J. Composite Materials*, 20(5):485–500.

42. Lie, B. T., S. T. Sun and L. Dahsin. 1987. "An Assessment of Damping Measurement in the Evaluation of Integrity of Composite Beams," *J. Reinforced Plastics and Composites*, 6:114–125.

43. Lin, C. H. 1989. "A Low Frequency Axial Oscillation Technique for Composite Material Damping Measurement," *J. Composite Materials*, 23(1):92–105.

44. Lin, D. X., R. G. Ni and R. D. Adams. 1984. "Prediction and Measurement of the Vibrational Damping Parameters of Carbon and Glass Fibre Reinforced Plastic Plates," *J. Composite Materials*, 18(2):132–152.

45. Malmeister, A. K., V. P. Tamuzh and G. A. Teters. 1980. *Strength of Polymer and Composite Materials* (in Russian). Riga: Zinatne.

46. Matveev, V. V., ed. 1985. *Vibration Damping of Deformable Solids* (in Russian). Kiev: Naukova Dumka.

47. Minomo, M. 1982. *Proceedings of the 13th Int. Symp. Space Technol. and Sci.*, Tokyo, June 28–July 3, 1982, pp. 367–372.

48. Moskvitin, V. V. 1984. *Cyclic Loading of Structure Elements* (in Russian). Moscow: Nauka.

49. Nashif, A. D., D. I. G. Johnes and J. P. Henderson, eds. 1985. *Vibration Damping*. New York: John Wiley & Sons, Inc.

50. Natarajan, R. T. and A. F. Lewis. 1976. "Advanced Composite Constrained Layer Laminates," *J. Composite Materials*, 10(4):220–230.

51. Nelson, D. J. and J. W. Hancock. 1978. "Interfacial Slip and Damping in Fibre Reinforced Composites," *J. Materials Science*, 13:2429–2440.

52. Ni, R. G. and R. D. Adams. 1984. "The Damping and Dynamic Moduli of Symmetric Laminated Composite Beams—Theoretical and Experimental Results," *J. Composite Materials*, 18(2):104–121.

53. Ni, R. G. and R. D. Adams. 1984. "A Rational Method for Obtaining the Dynamic Mechanical Properties of Laminae for Predicting the Stiffness and Damping of Laminated Plates and Beams," *Composites*, 15(3):193–199.

54. Ni, R. G., D. X. Lin and R. D. Adams. 1984. "The Dynamic Properties of Carbon-Glass Fibre Sandwich-Laminated Composites: Theoretical, Experimental and Economic Considerations," *Composites*, 15(4):297–304.

55. O'Donoghue, P. E. and S. N. Atlure. 1985. *Proceedings of the 26th Struct., Struct. Dyn. and Mater. Conf.*, Orlando, FL, April 15–17, 1985, pp. 31–42.

56. Palmov, V. A. 1976. *Vibrations of Elasto-Plastic Solids* (in Russian). Moscow: Nauka.

57. Panovko, Ya. G. 1960. *Internal Friction in Vibrating Elastic Systems* (in Russian). Moscow: Physmathgiz.

58. Paipetis, S. A. and P. Grootenhuis. 1979. "The Dynamic Properties of Fibre Reinforced Viscoelastic Composites," *Fibre Science and Technology* (12):353–376.

59. Paipetis, S. A. and P. Grootenhuis. 1979. "The Dynamic Properties of Particle Reinforced Viscoelastic Composites," *Fibre Science and Technology* (12):377–393.

60. Pelekh, B. L. and B. I. Salyakh, eds. 1990. *Experimental Methods for Investigating the Dynamic Behavior of Composite Structures.* Kiev: Naukova Dumka.

61. Pisarenko, G. S. 1985. *Generalized Non-Linear Model for Consideration of Energy Dissipation in Vibrations* (in Russian) Kiev: Naukova Dumka.

62. Pisarenko, G. S., V. V. Matveev and A. P. Yakovlev, eds. 1976. *Methods for Determining the Vibration Damping Behavior of Elastic Systems* (in Russian). Kiev: Naukova Dumka.

63. Plunkett, R. 1980. "Damping in Fiber Reinforced Laminated Composites at High Strain," *J. Composite Materials Supplement,* 14:109–117.

64. Plunkett, R. 1983. *Mechanics of Composite Materials: Recent Adv. Proc. IUTAM Symp.,* Blacksburg, VA, August 16–19, 1983, pp. 93–104.

65. Pobedrya, B. E., ed. 1984. *Mechanics of Composite Materials.* Moscow: Moscow University Publishing.

66. Reddy, J. N. 1979. "Free Vibration of Antisymmetric, Angle-Ply Laminated Plates Including Transverse Shear Deformation by the Finite Element Method," *J. Sound and Vibration,* 66(4):565–576.

67. Saravanos, D. A. and C. C. Chamis. 1990. "Mechanics of Damping for Fiber Composite Laminates Including Hygro-Thermal Effects," *AIAA Journal* (10):1813–1819.

68. Saravanos, D. A. and C. C. Chamis. 1990. "Unified Micromechanics of Damping for Unidirectional and Off-Axis Fiber Composites," *J. of Composites Technology and Research,* 12(1):31–40.

69. Sborovsky, A. K., Yu. A. Nikolsky and V. D. Popov. 1967. *Vibrations of the Ships with the Hulls of Glass Fiber Reinforced Plastic* (in Russian). Leningrad: Sudostroenie.

70. Schults, A. B. and S. W. Tsai. 1968. "Dynamic Moduli and Damping Relations in Fiber-Reinforced Composites," *J. Composite Materials,* 2(3):368–379.

71. Schults, A. B. and S. W. Tsai. 1969. "Measurement of Complex Dynamic Moduli for Laminated Fiber-Reinforced Composites," *J. Composite Materials,* 3(4):434–443.

72. Sharma, M. G. and C. S. Hong. 1975. "Structural Damping of Composites under Sustained Vibratory Stresses," *Polymer Engineering and Science,* 15(11):805–810.

73. Shepery, R. A. 1974. "Visco-Elastic Behavior of Composite Materials," in *Composite Materials, Vol. 2,* J. Lawrence, Broutman and R. H. Krock, eds.; in *Mechanics of Composite Materials,* G. P. Sendeckyj, ed., New York and London: Academic Press, pp. 102–194.

74. Silverman, E. M. and R. J. Jones. 1988. *Mater.-Pathway Future: 33rd Int. SAMPE Symp. and Exhib.,* Anaheim, CA, Covina, CA, March 7–10, 1988, pp. 1418–1432.

75. Sin, C. C. and C. W. Bert. 1974. "Sinusoidal Response of Composite-Material Plates with Material Damping," *J. Engineering for Industry* (5):603–610.

76. Spirnak, G. T. and J. R. Vinson. 1990. "The Effect of Temperature on the Material Damping of Graphite/Epoxy Composites in a Simulated Space Environment," *Trans. ASME J. Eng. Mater. and Technol.,* 112(3):277–279.

77. Suarez, S. A., R. F. Gibson and L. R. Deobald. 1984. "Random and Impulse Techniques for Measurement of Damping in Composite Materials," *Experimental Techniques*, 8(10):19–24.

78. Suarez, S. A., R. F. Gibson, C. T. Sun and S. K. Chaturvedi. 1986 "The Influence of Fiber Length and Fiber Orientation on Damping and Stiffness of Polymer Composite Materials," *Exp. Mech.*, 26(2):175–184.

79. Sun, C. T., S. K. Chaturvedi and R. F. Gibson. 1985. "Internal Damping of Short-Fiber Reinforced Polymer Matrix Composites," *Int. J. Computers & Structures*, 20(1–3): 391–400.

80. Sun, C. T., R. F. Gibson and S. K. Chaturvedi. 1985. "Internal Material Damping of Polymer Matrix Composites under Off-Axis Loading," *J. Materials Science*, 20(7): 2575–2585.

81. Thornton, P. H., J. J. Harwood and P. Beadmore. 1985. "Fiber Reinforced Plastic Composites for Energy Absorption Purposes," *Compos. Sci. and Technol.*, 24(4):257–298.

82. Timoshenko, S. P., ed. 1967. *Vibrations in Engineering* (in Russian). Moscow: Nauka.

83. Torvik, P. J. 1980. *Damping Application for Vibration Control, Meet. ASME*, Chicago, November 16–21, 1980, pp. 85–112.

84. Torvik, P. J. 1989. *Proceedings of AIAA/ASME/ASCE/AMS/ASC 30th Struct., Struct. Dyn. and Mater. Conf.*, Mobile, AL, April 3–5, 1989, Collect. Tech. Pap. Pt. 4, Washington, D.C., pp. 2246–2259.

85. Vanin, G. A., ed. 1985. *Micromechanics of Composite Materials* (in Russian). Kiev: Naukova Dumka.

86. 1984. "Vibration Damping Studied for Weapons in Space," *Aviation Week and Space Technology*, 121(2/VII):54–55.

87. Wang, S. F. and A. A. Ogale. 1989. "Influence of Aging on Transient and Dynamic Mechanical Properties of Carbon Fiber/Epoxy Composites," *SAMPE Quart.*, 20(2): 9–13.

88. White, R. G. and E. M. Y. Abdin. 1985. "Dynamic Properties of Aligned Short Carbon Fibre-Reinforced Plastics in Flexure and Torsion," *Composites*, 16(4):293–306.

89. Wilson, T. W., R. E. Fornes, R. D. Gilbert and J. D. Memory. 1988. "Effect of Ionizing Radiation on the Dynamic Mechanical Properties of an Epoxy and Graphite Fiber/Epoxy Composite," *J. Polym. Sci. B*, 26(10):2029–2042.

90. Wolfenden, A. and J. M. Wolla. 1989. "Mechanical Damping and Dynamic Modulus Measurements in Alumina and Tungsten Fibre-Reinforced Aluminium Composites," *J. Mater. Sci.*, 24(9):3205–3212.

91. Worth, R. A. 1986. "Properties of Glass-Fiber Reinforced Polypropylene Subjected to Pure Bending," *Polymer Engineering and Science*, 26(9):1293–1296.

92. Wright, G. C. 1972. "The Dynamic Properties of Glass and Carbon Fibre Reinforced Plastic Beams," *J. Sound and Vibration* (2):205–212.

93. Yakovlev, A. P., ed. 1985. *Dissipative Behavior of Non-Uniform Materials and Systems* (in Russian). Kiev: Naukova Dumka.

94. Zener, C. M., ed. 1968. *Elasticity and Inelasticity of Metals*. Chicago: Univ. of Chicago Press.

95. Zinoviev, P. A. and Yu. N. Ermakov. 1985. "Anisotropy of Fibrous Composites Dissipative Behavior," *Mekhanika Komposytnykh Materialov* (in Russian) (5):816–825.

96. Zinoviev, P. A. and Yu. N. Ermakov. 1985. "Dissipative Behavior of Unidirectional Fibrous Composite," *Izvestiya Vuzov* (in Russian) (12):16–21.

97. Zinoviev, P. A. and Yu. N. Ermakov. 1986. "Elasto-Dissipative Characteristic Matrices for Unidirectional Fibrous Composite. Coordinate Transformation," *Izvestiya Vuzov* (in Russian) (2):46–52.

98. Zinoviev, P. A. and Yu. N. Ermakov. 1986. "Energy Dissipation under Bending of Multilayered Fibrous Composites," *Izvestiya Vuzov* (in Russian) (4):15–20.

99. Zinoviev, P. A. and Yu. N. Ermakov. 1986. "Energy Dissipation of Fibrous Composite Bodies under Vibrations. Structural Model," *Application of Plastic Materials in Machine-Building. Proceedings of Moscow High Technical School* (in Russian) (21): 37–54.